# Image Understanding Using Sparse Representations

# Synthesis Lectures on Image, Video, and Multimedia Processing

Editor
**Alan C. Bovik,** *University of Texas, Austin*

The Lectures on Image, Video and Multimedia Processing are intended to provide a unique and groundbreaking forum for the world's experts in the field to express their knowledge in unique and effective ways. It is our intention that the Series will contain Lectures of basic, intermediate, and advanced material depending on the topical matter and the authors' level of discourse. It is also intended that these Lectures depart from the usual dry textbook format and instead give the author the opportunity to speak more directly to the reader, and to unfold the subject matter from a more personal point of view. The success of this candid approach to technical writing will rest on our selection of exceptionally distinguished authors, who have been chosen for their noteworthy leadership in developing new ideas in image, video, and multimedia processing research, development, and education.

In terms of the subject matter for the series, there are few limitations that we will impose other than the Lectures be related to aspects of the imaging sciences that are relevant to furthering our understanding of the processes by which images, videos, and multimedia signals are formed, processed for various tasks, and perceived by human viewers. These categories are naturally quite broad, for two reasons: First, measuring, processing, and understanding perceptual signals involves broad categories of scientific inquiry, including optics, surface physics, visual psychophysics and neurophysiology, information theory, computer graphics, display and printing technology, artificial intelligence, neural networks, harmonic analysis, and so on. Secondly, the domain of application of these methods is limited only by the number of branches of science, engineering, and industry that utilize audio, visual, and other perceptual signals to convey information. We anticipate that the Lectures in this series will dramatically influence future thought on these subjects as the Twenty-First Century unfolds.

Image Understanding Using Sparse Representations
Jayaraman J. Thiagarajan, Karthikeyan Natesan Ramamurthy, Pavan Turaga, and Andreas Spanias
2014

Contextual Analysis of Videos
Myo Thida, How-lung Eng, Dorothy Monekosso, and Paolo Remagnino
2013

Wavelet Image Compression
William A. Pearlman
2013

Remote Sensing Image Processing
Gustavo Camps-Valls, Devis Tuia, Luis Gómez-Chova, Sandra Jiménez, and Jesús Malo
2011

The Structure and Properties of Color Spaces and the Representation of Color Images
Eric Dubois
2009

Biomedical Image Analysis: Segmentation
Scott T. Acton and Nilanjan Ray
2009

Joint Source-Channel Video Transmission
Fan Zhai and Aggelos Katsaggelos
2007

Super Resolution of Images and Video
Aggelos K. Katsaggelos, Rafael Molina, and Javier Mateos
2007

Tensor Voting: A Perceptual Organization Approach to Computer Vision and Machine Learning
Philippos Mordohai and Gérard Medioni
2006

Light Field Sampling
Cha Zhang and Tsuhan Chen
2006

Real-Time Image and Video Processing: From Research to Reality
Nasser Kehtarnavaz and Mark Gamadia
2006

MPEG-4 Beyond Conventional Video Coding: Object Coding, Resilience, and Scalability
Mihaela van der Schaar, Deepak S Turaga, and Thomas Stockhammer
2006

Modern Image Quality Assessment
Zhou Wang and Alan C. Bovik
2006

Biomedical Image Analysis: Tracking
Scott T. Acton and Nilanjan Ray
2006

Recognition of Humans and Their Activities Using Video
Rama Chellappa, Amit K. Roy-Chowdhury, and S. Kevin Zhou
2005

Image Understanding Using Sparse Representations

Jayaraman J. Thiagarajan, Karthikeyan Natesan Ramamurthy, Pavan Turaga, and Andreas Spanias

ISBN: 978-3-031-01122-1     paperback
ISBN: 978-3-031-02250-0     ebook

DOI 10.1007/978-3-031-02250-0

A Publication in the Springer series
*SYNTHESIS LECTURES ON IMAGE, VIDEO, AND MULTIMEDIA PROCESSING*

Lecture #15
Series Editor: Alan C. Bovik, *University of Texas, Austin*
Series ISSN
Synthesis Lectures on Image, Video, and Multimedia Processing
Print 1559-8136   Electronic 1559-8144

# Image Understanding
# Using Sparse Representations

Jayaraman J. Thiagarajan
Lawrence Livermore National Laboratory

Karthikeyan Natesan Ramamurthy
IBM Thomas J. Watson Research Center

Pavan Turaga
Arizona State University

Andreas Spanias
Arizona State University

*SYNTHESIS LECTURES ON IMAGE, VIDEO, AND MULTIMEDIA PROCESSING #15*

# ABSTRACT

Image understanding has been playing an increasingly crucial role in several inverse problems and computer vision. Sparse models form an important component in image understanding, since they emulate the activity of neural receptors in the primary visual cortex of the human brain. Sparse methods have been utilized in several learning problems because of their ability to provide parsimonious, interpretable, and efficient models. Exploiting the sparsity of natural signals has led to advances in several application areas including image compression, denoising, inpainting, compressed sensing, blind source separation, super-resolution, and classification.

The primary goal of this book is to present the theory and algorithmic considerations in using sparse models for image understanding and computer vision applications. To this end, algorithms for obtaining sparse representations and their performance guarantees are discussed in the initial chapters. Furthermore, approaches for designing overcomplete, data-adapted dictionaries to model natural images are described. The development of theory behind dictionary learning involves exploring its connection to unsupervised clustering and analyzing its generalization characteristics using principles from statistical learning theory. An exciting application area that has benefited extensively from the theory of sparse representations is compressed sensing of image and video data. Theory and algorithms pertinent to measurement design, recovery, and model-based compressed sensing are presented. The paradigm of sparse models, when suitably integrated with powerful machine learning frameworks, can lead to advances in computer vision applications such as object recognition, clustering, segmentation, and activity recognition. Frameworks that enhance the performance of sparse models in such applications by imposing constraints based on the prior discriminatory information and the underlying geometrical structure, and kernelizing the sparse coding and dictionary learning methods are presented. In addition to presenting theoretical fundamentals in sparse learning, this book provides a platform for interested readers to explore the vastly growing application domains of sparse representations.

# KEYWORDS

sparse representations, natural images, image reconstruction, image recovery, image classification, dictionary learning, clustering, compressed sensing, kernel methods, graph embedding

# Contents

**1 Introduction** .................................................. 1
  1.1 Modeling Natural Images ..................................... 1
  1.2 Natural Image Statistics ...................................... 1
  1.3 Sparseness in Biological Vision ............................... 3
  1.4 The Generative Model for Sparse Coding ....................... 5
  1.5 Sparse Models for Image Reconstruction ....................... 6
     1.5.1 Dictionary Design ....................................... 6
     1.5.2 Example Applications .................................... 7
  1.6 Sparse Models for Recognition ............................... 9
     1.6.1 Discriminative Dictionaries ............................. 10
     1.6.2 Bag of Words and its Generalizations .................... 11
     1.6.3 Dictionary Design with Graph Embedding Constraints ...... 12
     1.6.4 Kernel Sparse Methods .................................. 12

**2 Sparse Representations** ....................................... 13
  2.1 The Sparsity Regularization ................................. 13
     2.1.1 Other Sparsity Regularizations .......................... 14
     2.1.2 Non-Negative Sparse Representations ..................... 16
  2.2 Geometrical Interpretation .................................. 17
  2.3 Uniqueness of $\ell_0$ and its Equivalence to the $\ell_1$ Solution ..... 17
     2.3.1 Phase Transitions ....................................... 20
  2.4 Numerical Methods for Sparse Coding ......................... 20
     2.4.1 Optimality conditions ................................... 21
     2.4.2 Basis Pursuit ........................................... 22
     2.4.3 Greedy Pursuit Methods ................................. 23
     2.4.4 Feature-Sign Search ..................................... 26
     2.4.5 Iterated Shrinkage Methods .............................. 28

**3 Dictionary Learning: Theory and Algorithms** .................. 29
  3.1 Dictionary Learning and Clustering .......................... 31
     3.1.1 Clustering Procedures ................................... 31

         3.1.2  Probabilistic Formulation . . . . . . . . . . . . . . . . . . . . . . . . . . . . . . 32
3.2  Learning Algorithms . . . . . . . . . . . . . . . . . . . . . . . . . . . . . . . . . . . . . . . . 33
         3.2.1  Method of Optimal Directions . . . . . . . . . . . . . . . . . . . . . . . . . . . . 33
         3.2.2  K-SVD . . . . . . . . . . . . . . . . . . . . . . . . . . . . . . . . . . . . . . . . . . . . . . . 34
         3.2.3  Multilevel Dictionaries . . . . . . . . . . . . . . . . . . . . . . . . . . . . . . . . . 36
         3.2.4  Online Dictionary Learning . . . . . . . . . . . . . . . . . . . . . . . . . . . . . 41
         3.2.5  Learning Structured Sparse Models . . . . . . . . . . . . . . . . . . . . . . 42
         3.2.6  Sparse Coding Using Examples . . . . . . . . . . . . . . . . . . . . . . . . . . 46
3.3  Stability and Generalizability of Learned Dictionaries . . . . . . . . . . . . . 50
         3.3.1  Empirical Risk Minimization . . . . . . . . . . . . . . . . . . . . . . . . . . . . 51
         3.3.2  An Example Case: Multilevel Dictionary Learning . . . . . . . . . . 52

**4  Compressed Sensing** . . . . . . . . . . . . . . . . . . . . . . . . . . . . . . . . . . . . . . . . . **55**
4.1  Measurement Matrix Design . . . . . . . . . . . . . . . . . . . . . . . . . . . . . . . . . . 56
         4.1.1  The Restricted Isometry Property . . . . . . . . . . . . . . . . . . . . . . . . 56
         4.1.2  Geometric Interpretation . . . . . . . . . . . . . . . . . . . . . . . . . . . . . . . 58
         4.1.3  Optimized Measurements . . . . . . . . . . . . . . . . . . . . . . . . . . . . . . . 58
4.2  Compressive Sensing of Natural Images . . . . . . . . . . . . . . . . . . . . . . . . . 59
4.3  Video Compressive Sensing . . . . . . . . . . . . . . . . . . . . . . . . . . . . . . . . . . . 60
         4.3.1  Frame-by-Frame Compressive Recovery . . . . . . . . . . . . . . . . . . 61
         4.3.2  Model-Based Video Compressive Sensing . . . . . . . . . . . . . . . . . 62
         4.3.3  Direct Feature Extraction from Compressed Videos . . . . . . . . . 63

**5  Sparse Models in Recognition** . . . . . . . . . . . . . . . . . . . . . . . . . . . . . . . . **67**
5.1  A Simple Classification Setup . . . . . . . . . . . . . . . . . . . . . . . . . . . . . . . . . 67
5.2  Discriminative Dictionary Learning . . . . . . . . . . . . . . . . . . . . . . . . . . . . 71
5.3  Sparse-Coding-Based Subspace Identification . . . . . . . . . . . . . . . . . . . 72
5.4  Using Unlabeled Data in Supervised Learning . . . . . . . . . . . . . . . . . . . 73
5.5  Generalizing Spatial Pyramids . . . . . . . . . . . . . . . . . . . . . . . . . . . . . . . . 74
         5.5.1  Supervised Dictionary Optimization . . . . . . . . . . . . . . . . . . . . . 77
5.6  Locality in Sparse Models . . . . . . . . . . . . . . . . . . . . . . . . . . . . . . . . . . . . 78
         5.6.1  Local Sparse Coding . . . . . . . . . . . . . . . . . . . . . . . . . . . . . . . . . . . 78
         5.6.2  Dictionary Design . . . . . . . . . . . . . . . . . . . . . . . . . . . . . . . . . . . . . 79
5.7  Incorporating Graph Embedding Constraints . . . . . . . . . . . . . . . . . . . . 80
         5.7.1  Laplacian Sparse Coding . . . . . . . . . . . . . . . . . . . . . . . . . . . . . . . 81
         5.7.2  Local Discriminant Sparse Coding . . . . . . . . . . . . . . . . . . . . . . . 81
5.8  Kernel Methods in Sparse Coding . . . . . . . . . . . . . . . . . . . . . . . . . . . . . 83

5.8.1 Kernel Sparse Representations .................................. 84

5.8.2 Kernel Dictionaries in Representation and Discrimination .......... 85

5.8.3 Combining Diverse Features.................................... 87

5.8.4 Application: Tumor Identification ............................. 89

**Bibliography** ......................................................... **91**

**Authors' Biographies** ............................................... **105**

# CHAPTER 1

# Introduction

## 1.1  MODELING NATURAL IMAGES

Natural images are pervasive entities that have been studied over the past five decades. The term "natural images" includes those images that are usually present in the environment in which we live. The definition encompasses a general class that includes scenes and objects present in nature as well as man-made entities such as cars, buildings, and so on. Natural images have interested scientists from a wide range of fields from psychology to applied mathematics. Though these images have rich variability when viewed as a whole, their local regions often demonstrate substantial similarities. Some examples of natural images that we have used in this book are given in Figure 1.1.

Many proposed models for representing natural images have tried to mimic the processing of images by the human visual system. Generally, image representation is concerned with low-level vision. Hence, using local regions or *patches* of images, and exploiting the local similarities in order to build a set of features, can be beneficial in image representation. The representative features that are extracted from local image regions are referred to as *low-level features*. It is imperative that we understand and incorporate our knowledge of the human visual processing principles, as well as the local image statistics, when building systems for representating images. As a result, exploring natural image statistics has become an integral part in the design of efficient image representation systems.

As we shift our focus from the problem of representing images to the task of recognizing objects or classifying scenes, it can easily be seen that it cannot be performed just by representing the local regions of images using the low-level features. Considering the problem of recognizing an object from a scene, it can be seen that the local features extracted will not be invariant to global transformations of the object, its local deformations, and occlusion. Hence, we need to ensure that they capture the key invariants of an object to transformations and, to some extent, occlusions, instead of just extracting low-level features that can provide a good representation. Such features can be used to build image descriptors that are helpful in recognition and classification.

## 1.2  NATURAL IMAGE STATISTICS

A digital image denoted using the symbol $I$ is represented using pixel values between 0 and 255, for grayscale. Image statistics refer to the various statistical information of the raw pixel values, or their transformed versions. By incorporating natural image statistics as prior information in

**Figure 1.1:** Examples of some natural images used in this book. These images are obtained from the Berkeley segmentation dataset (BSDS) [1].

a Bayesian inference scheme, one can hope to learn features that are adapted to the image data, rather than try to use classical transform methods that may not be adapted well to the data.

Since analyzing raw pixel statistics hardly produces any useful information, analysis of natural image statistics has been performed in Fourier [2] and Wavelet domains [3]. It is observed that the power spectrum changes as $1/f^2$, where $f$ is the frequency, and this has been one of the earliest proofs of redundancy in natural images. The marginal statistics of wavelet coefficients in a subband, illustrated in Figure 1.2, show that the distributions are more peaky and heavily tailed when compared to a Gaussian distribution. Furthermore, the joint statistics would indicate a strong dependence between the coefficients in different subbands. The coefficient statistics obtained using log-Gabor filters on natural images also reveal a peaked distribution. Figure 1.3 shows the output distribution at scale 1 and orientation 1 for the *Barbara* image. These log-Gabor filters are typically used to detect edges in natural images. The histogram of the coefficients implies that most of the responses obtained from the filters are close to zero. This demonstration shows that natural images can be sparsely decomposed in any set of basis functions that can identify "edges" in the images efficiently.

Besides understanding the statistics, analyzing the topological structure of the natural image patch distribution can also provide some crucial insights. As a result of such an effort reported in [4, 5], it has been shown that high-contrast image patches lie in clusters and near low-dimensional manifolds. Patches of size $3 \times 3$ were chosen from natural images and preprocessed in the logarithmic scale to remove the mean and normalize the contrast values. They were then projected using the discrete cosine transform (DCT) basis and normalized to unit norm, which makes the processed data lie in a 7−sphere in $\mathbb{R}^8$. By computing the fraction of the data points that are near the dense sampling of the surface of the sphere, the empirical distribution of the data can be determined. It has been observed that a majority of the data points are concentrated in a few high-density regions of the sphere. The sample points corresponding to the high-density regions are similar to blurred step edges for natural image patches. Further topological analysis of natural image patches has been performed in [5], which suggests that features that can efficiently represent a large portion of natural images can be extracted by computing topological equivalents to the space of natural image patches.

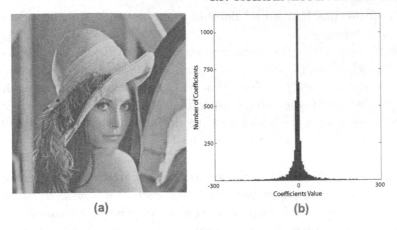

**Figure 1.2:** (a) The *Lena* image, (b) marginal statistics of wavelet coefficients in a subband.

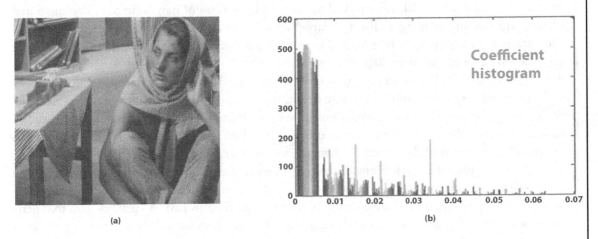

**Figure 1.3:** (a) The *Barbara* image, (b) log-Gabor coefficient statistics at scale 1 and orientation 1 for the image.

## 1.3   SPARSENESS IN BIOLOGICAL VISION

Most of our cognitive functions and perceptual processes are carried out in the neocortex, which is the largest part of the human brain. The primary visual cortex, also referred to as V1, is the part of the neocortex that receives visual input from the retina. The Nobel Prize-winning discoveries of Hubel and Weisel showed that the primary visual cortex consists of cells responsive to simple and complex features in the input. V1 has receptive fields that are characterized as being spatially localized, oriented, and bandpass. In other words, they are selective to the structure of the visual input at different spatial scales. One approach to understanding the response properties of visual neurons has been to consider their relationship to the statistical structure of natural images in terms of efficient coding.

A generative model that constructs random noise intensities typically results in images with equal probability. However, the statistics of natural image patches have been shown to contain a variety of regularities. Hence, the probability of generating a natural image using the random noise model is extremely low. In other words, the redundancy in natural images makes the evolution of an efficient visual system possible. Extending this argument, understanding the behavior of the visual system in exploiting redundancy will enable us to build a plausible explanation for the coding behavior of visual neurons [6].

The optimality of the visual coding process can be addressed with respect to different metrics of efficiency. One of the most commonly used metrics is *representation efficiency*. Several approaches have been developed to explore the properties of neurons involved in image representation. In [7], it was first reported that the neurons found in the V1 showed similarities to Gabor filters, and, hence, different computational models were developed based on this relation [8, 9]. Furthermore, the $1/f^k$ fall-off observed in the Fourier spectra of natural images demonstrated the redundancy in the images. The $1/f$ structure arises because of the pairwise correlations in the data and the observation that natural images are approximately scale invariant. The pairwise correlations account for about 40% of the total redundancy in natural scenes. Any representation with significant correlations implies that most of the signal lies in a subspace within the larger space of possible representations [10] and, hence, data can be represented with much-reduced dimensionality. Though pairwise correlations have been an important form of redundancy, studies have shown that there exists a number of other forms. Two images with similar $1/f$ spectra can be described in terms of differences in their sparse structure.

For example, all linear representations with a noise image having a $1/f$ spectra will result in a Gaussian response distribution. However, with natural images, the histogram of the responses will be non-Gaussian for a suitable choice of linear filters (Figure 1.3). A visual system that generates such response distributions with high kurtosis (fourth statistical moment) can produce a representation with visual neurons that are maximally independent. In other words, codes with maximal independence will activate neurons with maximal unique information. With respect to biological vision, *sparsity* implies that a small proportion of the neurons are active, and the rest of the population is inactive with high probability. Olshausen and Field showed that, by designing a neural network that attempts to find sparse linear codes for natural scenes, we can obtain a family of localized, oriented, and bandpass basis functions, similar to those found in the primary visual cortex. This evidenced that, at the level of V1, visual system representations can efficiently match the sparseness of natural scenes. However, it must be emphasized that the sparse outputs of these models result from the sparse structure of the data. For example, similar arguments cannot be made for white-noise images. Further studies have shown that the visual neurons produce sparse responses in higher stages of cortical processing, such as inferotemporal cortex, in addition to the primary visual cortex. However, there is no biological evidence to show that these sparse responses imply efficient representation of the environment by the neurons.

Measuring the efficiency of neural systems is very complicated compared to that of engineered visual systems. In addition to characterizing visual representations, there is a need to understand its dependency on efficient learning and development. A learning algorithm must address a number of problems pertinent to generalization. For example, invariance is an important property that the learning algorithm must possess. In this case, efficient learning is determined by its ability to balance the selection of suitable neurons and achieving invariance across examples with features that vary. In other words, it is critical for the algorithm to generalize to multiple instances of an object in the images. At one end, this problem can be addressed by building a neuron for every object. Simpler tasks, such as distinguishing between different faces, can be efficiently performed using this approach. However, the number of object detectors in such a system would be exorbitantly high. On the other end, sparse codes with more neurons are necessary for visually challenging tasks, such as object categorization. Hence, the general task of object recognition requires different strategies with varying degrees of sparseness.

Another important property of the representations in biological vision is that they are highly overcomplete, and, hence, they involve significant redundancy. This is one of the motivating factors behind using overcompleteness as an efficient way to model the redundancy in images [11]. Furthermore, overcompleteness can result in highly sparse representations, when compared to using complete codes, and this property is very crucial for generalization during learning. Though a comprehensive theory to describe the human visual processing has not yet been developed, it has been found that considering different measures of efficiency together is a crucial step toward this direction.

## 1.4 THE GENERATIVE MODEL FOR SPARSE CODING

The linear generative model for sparse representation of an image patch $\mathbf{x}$ is given by

$$\mathbf{x} = \mathbf{D}\mathbf{a}, \tag{1.1}$$

where $\mathbf{x} \in \mathbb{R}^M$ is the arbitrary image patch, $\mathbf{D} \in \mathbb{R}^{M \times K}$ is the set of $K$ elementary features, and $\mathbf{a} \in \mathbb{R}^K$ is the coefficient vector. If we assume that the coefficient vector is sparse and has statistically independent components, the elements of the dictionary $\mathbf{D}$ can be inferred from the generative model using appropriate constraints. The reason for assuming a sparse prior on the elements of $\mathbf{a}$ is the intuition that images allow for their efficient representation as a sparse linear combination of patterns, such as edges, lines, and other elementary features [12]. Let us assume that all the $T$ patches extracted from the image $I$ are given by the matrix $\mathbf{X} \in \mathbb{R}^{M \times T}$, and the coefficient vectors for all the $T$ patches are given by the matrix $\mathbf{A} \in \mathbb{R}^{K \times T}$. The likelihood for the image $I$, which is represented as the matrix $\mathbf{X}$, is given by

$$\log p(\mathbf{X}|\mathbf{D}) = \int p(\mathbf{X}|\mathbf{D}, \mathbf{A}) p(\mathbf{A}) d\mathbf{A}. \tag{1.2}$$

$p(\mathbf{X}|\mathbf{D}, \mathbf{A})$ can be computed using the linear generative model assumed in (1.1) and $p(\mathbf{A})$ is the prior that enforces sparsity constraints on the entries of the coefficient vectors. The dictionary now can be inferred as

$$\hat{\mathbf{D}} = \underset{\mathbf{D}}{\mathrm{argmax}} \log p(\mathbf{X}|\mathbf{D}). \qquad (1.3)$$

The strategy proposed by Olshausen and Field [13] to determine the dictionary $\mathbf{D}$ is to use the generative model (1.1) and an approximation of (1.3), along with the constraint that the columns of $\mathbf{D}$ are of unit $\ell_2$ norm. The dictionary $\mathbf{D}$ here is overcomplete, i.e., the number of dictionary elements is greater than the dimensionality of the image patch, $K > M$. This leads to infinite solutions for the coefficient vector in (1.1) and, hence, the assumption on the sparsity of $\mathbf{a}$ can be used to choose the most appropriate dictionary elements that represent $\mathbf{x}$. Our strategy for computing sparse representations and learning dictionaries will also be based on this generative model. Although the generative model assumes that the distribution of coefficients is independent, there will still be statistical dependencies between the inferred coefficients. One of the reasons for this could be that the elementary features corresponding to the non-zero coefficients may occlude each other partially. In addition, the features themselves could co-occur to represent a more complex pattern in a patch.

It should be noted that using a model such as (1.1), along with sparsity constraints on $\mathbf{a}$, deviates from the linear representation framework. Assume that we have a representation that is $S-$sparse, (i.e.), only $S$ out of $K$ coefficients are non-zero at any time in $\mathbf{a}$. There are totally $\binom{K}{S}$ such combinations possible, each representing an $S-$dimensional subspace assuming that each set of $S$ chosen dictionary elements are independent. This model leads to a union of $S-$dimensional subspaces where the signal to be represented can possibly reside. It is clearly not a linear model, as the sum of signals lying in two $S-$dimensional subspaces can possibly lie in a $2S-$dimensional subspace.

## 1.5　SPARSE MODELS FOR IMAGE RECONSTRUCTION

Images can be modeled using sparse representations on predefined dictionaries as well as on those learned from the data itself. Predefined dictionaries obtained from the discrete cosine transform (DCT), wavelet, wedgelet, and curvelet transforms are widely used in signal and image processing applications. Since a single predefined dictionary cannot completely represent the patterns in an image, using a combination of predefined dictionaries is helpful in some applications.

### 1.5.1　DICTIONARY DESIGN

DCT dictionaries were some of the earliest dictionaries for patch-based image representation and are still used for initializing various dictionary learning procedures. Overcomplete DCT dictionaries can easily be constructed and have been used with sparse representations, performing substantially better than orthonormal DCT dictionaries. The set of predefined dictionaries that have probably found the widest applications in multiscale image representation are the wavelet

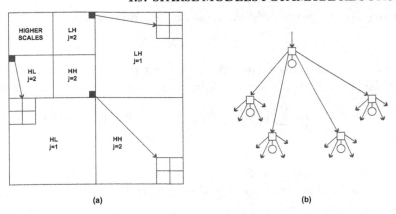

**Figure 1.4:** (a) Parent-child relationship between the wavelet coefficients across different scales, (b) the quad-tree model.

dictionaries. Wavelet coefficients have strong inter- and intra-scale dependencies and modeling them can be helpful in applications that exploit this additional redundancy. Many successful image coders consider either implicit or explicit statistical dependencies between the wavelet coefficients. For example, the JPEG-2000 standard [14] considers the neighboring coefficients in the adjacent subbands of the same scale jointly while coding. Figure 1.4 provides the quad-tree model for a parent-child relationship between the wavelet coefficients, and this structured sparsity has been used in denoising [15], pattern recognition using templates [16], and model-based compressive sensing frameworks [17, 18]. Such a tree-based model had been used in one of the earliest wavelet coders—the Embedded Zerotree Wavelet coder [19].

Using sparsity models with learned dictionaries has been very successful. The simplest model that assumes the coefficients are independent gives rise to the K-SVD learning algorithm [20]. If the sparsity patterns appear in groups or blocks, block-based sparse representations can be performed [21]. Additional structure can be imposed on group sparsity and each sparse pattern can be penalized accordingly. This gives rise to the structured sparsity frameworks, an area of ongoing research [22]. Apart from considering a structure when computing the coefficients, structured dictionaries can also be learned. An important class of dictionary learning algorithms imposes a hierarchical structure on the dictionary. For example, the multilevel dictionary learning [23] algorithm exploits the energy hierarchy found in natural images to learn dictionaries using multiple levels of 1—sparse representations.

## 1.5.2  EXAMPLE APPLICATIONS

Sparsity of the coefficients can be exploited in a variety of signal and image processing applications. Applications that use sparse models for recovery problems assume that the image data can be represented as a sparse linear combination of elements from an appropriate dictionary.

- **Denoising**: An image with additive noise can be represented as

$$\mathbf{X}_n = \mathbf{X} + \mathbf{N}, \tag{1.4}$$

where $\mathbf{X}_n$ is the noisy image (vectorized) and $\mathbf{N}$ represents the noise added to the image. The denoising problem that recovers $\mathbf{X}$ from its noisy counterpart can be posed as [24]

$$\{\hat{\mathbf{a}}_{ij}, \hat{\mathbf{X}}\} = \underset{\mathbf{a}_{ij}, \mathbf{X}}{\operatorname{argmin}} \sum_{ij} \|\mathbf{D}\mathbf{a}_{ij} - \mathbf{x}_{ij}\|, \text{ subj. to } \|\mathbf{a}_{ij}\|_0 \le S, \|\mathbf{X}_n - \mathbf{X}\|_2 \le \epsilon. \tag{1.5}$$

Here, $\mathbf{x}_{ij}$ and $\mathbf{a}_{ij}$ denote the patch at location $i, j$ in the image, and its corresponding coefficient vector, $\|.\|_0$ represents the $\ell_0$ norm, which counts the number of nonzero entries in a vector, and $\epsilon$ denotes the error goal that depends on the additive noise. This general model can be used for any predefined or learned dictionary that operates patchwise in the image. However, it is clear that the model can be improved by considering dependencies between the coefficients within a patch or across the patches themselves [25].

- **Image Inpainting**: Image inpainting is a problem where pixel values at some locations of the image would be unknown and have to be predicted. Assuming the simple case where the values of random pixels in a patch is unknown, we can express the incomplete patch as

$$\mathbf{z} = \boldsymbol{\Phi}\mathbf{x} + \mathbf{n}, \tag{1.6}$$

where $\mathbf{z} \in \mathbb{R}^N$ is the incomplete patch with unknown entries from $\mathbf{x}$ removed and $\boldsymbol{\Phi} \in \mathbb{R}^{N \times M}$ is an identity matrix with rows corresponding to unknown pixel values removed. Since we know that $\mathbf{x}$ is sparsely representable, we can consider

$$\mathbf{z} = \boldsymbol{\Phi}\mathbf{D}\mathbf{a} + \mathbf{n}, \tag{1.7}$$

which implies that $\mathbf{z}$ is sparsely representable using the equivalent dictionary $\boldsymbol{\Phi}\mathbf{D}$.

- **Compressive Recovery**: In compressive sensing, we *sense* the data $\mathbf{x}$ using a linear measurement system $\boldsymbol{\Phi} \in \mathbb{R}^{N \times M}$, where $N < M$. The linear measurement matrix usually consists of entries from independent entries from a Gaussian or Bernoulli distribution [26], although designing deterministic measurement systems is also possible [27]. This can be described by an equation similar to (1.6) and (1.7), though the matrix $\boldsymbol{\Phi}$ here is different. Recovery is performed by computing a sparse representation using the equivalent dictionary. It is also possible to design the linear measurement system optimized to the dictionary so that the recovery performance improves [28].

Note that for all three inverse problems discussed, i.e., denoising, inpainting, and compressive sensing, recovery performance depends on the sparsity of the uncorrupted/complete patch $\mathbf{y}$ and the dictionary used for recovery. For denoising, this will be the sparsifying dictionary $\mathbf{D}$, whereas, for compressive sensing and inpainting, this will be the equivalent dictionary $\boldsymbol{\Phi}\mathbf{D}$. The conditions on sparsity and the dictionary for recovery of a unique representation are discussed in Chapter 2.

- **Source Separation**: The source separation problem is different in spirit from the inverse problems discussed earlier. Assume that we have $K$ sparse sources given by the rows of $\mathbf{A} \in \mathbb{R}^{K \times T}$ and the mixing matrix (dictionary) given by $\mathbf{D}$. Note that the mixing matrix is usually overcomplete. The observations are noisy and mixed versions from the sources. The goal is to estimate the sources based on assumptions of their sparsity. While the inverse problems discussed above may or may not involve learning a dictionary, source separation requires that both the mixing matrix and the sparse sources be inferred from the observations.

## 1.6 SPARSE MODELS FOR RECOGNITION

The relevance of sparsity in machine learning systems has been studied extensively over the last decade. Deep Belief Networks (DBNs) have been used to effectively infer the data distribution and extract features in an unsupervised fashion [29]. Imposing sparsity constraints on DBNs for natural images will result in features that closely resemble their biological counterparts [30]. Sparse representations have also been known more likely to be separable in high-dimensional spaces [31], and, hence, helpful in classification tasks. However, empirical studies designed to evaluate the actual importance of sparsity in image classification state the contrary.

The authors of [32] argue that no performance improvement is gained by imposing sparsity in the feature extraction process, when this sparsity is not tailored for discrimination. Before discussing the discriminative models using sparse representations, it is imperative to understand why sparse representations computed without any additional constraints cannot be directly used for discrimination. It can be understood by considering a simple analogy. Consider a set of low-level features needed to represent a human face. It can easily be seen that a general set of features can efficiently represent a human face. However, if we try to classify male and female faces, for instance, directly using those features will result in a poor performance. For good classification performance, we may need to incorporate features that describe the discriminating characteristics between male and female faces, such as presence of facial hair, thickness of eyebrows, etc.

Computer vision frameworks that use sparse representations can be divided into two broad categories: (a) those that incorporate explicit discriminative constraints between classes when learning dictionaries or computing sparse codes; and (b) those that use the sparse codes as low-level features to generalize bag-of-words-based models for use in classification. As a first example, let us consider a standard object recognition architecture based on sparse features, similar to the ones developed in [33, 34]. Given a known set of filters (dictionary atoms), sparse features are extracted by coding local regions in an image. Note that, in some cases, it may be possible to replace the computationally intensive sparse coding process by a simple convolution of the patches with the linear filters, for a small or no loss in performance. The image filters can be learned using dictionary learning procedures, or can be pre-determined based on knowledge of the data. The architecture illustrated in figure 1.5 also includes a pre-processing step where operations such as conversion of color images to grayscale, whitening, and dimensionality reduction can be carried out.

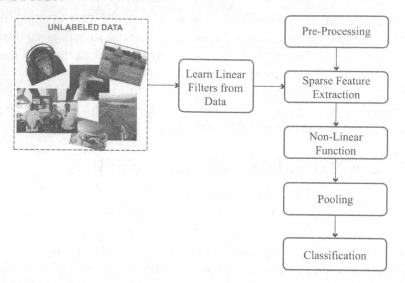

**Figure 1.5:** A biologically inspired standard object recognition architecture based on sparse representations.

Sparse feature extraction is followed by the application of a non-linear function, such as taking the absolute values of the coefficients, and pooling. Note that this procedure is common in biologically inspired multi-layer architectures for recognition. For example, the models proposed in [35] start with grayscale pixels and successively compute the "S" (simple) and the "C" (complex) layers in an alternating manner. The simple layers typically apply local filters to compute higher-order features by combining various lower-order features in the previous layer. On the other hand, the complex layers attempt to achieve invariance to minor pose and appearance changes by pooling features of the same unit in the previous layer. Since the dimension of the descriptors is too high for practical applications, downsampling is usually performed. Some of the commonly used pooling functions include Gaussian pooling, average pooling, and maximum-value pooling (*max-pooling*) in the neighborhood. The final step in the architecture is to learn a classifier that discriminates the pooled features belonging to the different classes.

## 1.6.1  DISCRIMINATIVE DICTIONARIES

The general approaches in using sparse representations for discrimination applications involve either learning class-specific dictionaries or a single dictionary for multiple classes. Class-specific dictionaries are learned for $C$ classes of the training data with the constraint that $\|\mathbf{x} - \mathbf{D}_c \mathbf{a}_c\|_2^2 <$ $\|\mathbf{x} - \mathbf{D}_j \mathbf{a}_j\|_2^2$. Here, $\mathbf{x}$ is the sample that belongs to class $c$, $\mathbf{D}_c$ and $\mathbf{D}_j$ are dictionaries that belong to classes $c$ and $j$ respectively, where $j \neq c$, and $\mathbf{a}_c$ and $\mathbf{a}_j$ are the sparse codes of $\mathbf{x}$ corresponding to those dictionaries. This constraint may be incorporated implicitly or explicitly. Classification is performed based on computing the representation error for test data with respect to all the dictionaries and choosing the class which yields the least representation error.

When a single dictionary is learned for multiple classes, classification is performed by incorporating some discrimination constraint directly in learning [36]. This can be given as

$$\min_{\mathbf{D}, \{\mathbf{a}_i\}, \boldsymbol{\theta}} \sum_i f(\mathbf{D}, \mathbf{a}_i, y_i, \boldsymbol{\theta}) + \lambda_1 \|\mathbf{x}_i - \mathbf{D}\mathbf{a}_i\|_2^2 \text{ subj. to } \|\mathbf{a}_i\|_0 \leq S, \tag{1.8}$$

where $\boldsymbol{\theta}$ are the parameters of the classifier that are optimized along with reducing the representation error, and $f(\mathbf{D}, \mathbf{a}_i, y_i, \boldsymbol{\theta})$ is the discrimination constraint that depends on the label $y_i$ of the training data $\mathbf{x}_i$. Although (1.8) incorporates the classifier parameters directly, it is also possible to implement alternative minimization schemes that iterate between optimizing the classifier model parameters and the dictionary. If data samples belonging to a certain class can be represented well using a sparse linear combination of data belonging to the same class, such as in the case of face images, classification can be performed by directly using the training data as a dictionary and comparing the relative projection errors of test data on different classes [37].

## 1.6.2  BAG OF WORDS AND ITS GENERALIZATIONS

Another approach for building recognition frameworks is to use sparse codes of local regions of an image as low-level features in order to build image specific features that can be used for training a classifier, such as a support vector machine (SVM). Since low-level features are common across all natural images, unlabeled images can be used to learn them and the labeled images can then be represented using the learned features. Finally, an image-specific feature can be derived by aggregating the low-level features and choosing the coefficient of the strongest feature, an operation referred to as max-pooling. The self-taught transfer learning approach described in [38] uses sparse representation to build such image-specific features.

Though low-level patch-based representation features are helpful in classification, inherently they are more suitable for efficient representation. Hence, it is more intuitive to construct discriminative features, even at the patch level, that are helpful in classification. The local sparse coding technique incorporates the locality constraint when computing the coefficient vector. This ensures that the active dictionary elements are close to the data represented by the sparse code. This has been shown to improve the discriminative power of the sparse code [39]. Another method of making the sparse code discriminative is to derive non-linear features, such as the scale-invariant feature transform (SIFT) and the histogram of gradients (HOG) at the patch level, and use them to compute sparse representations. The bag-of-words approach uses SIFT features at patch level and computes the image-level features by counting the number of times a feature is used in the image. A similar method is used by the spatial-pyramid matching framework, albeit at different spatial resolutions of the image [40]. Approaches based on sparse representations with local sparse codes, or sparse codes of non-linear patch-level features, use max-pooling in order to compute image-level features [39]. Max-pooling, by virtue of picking the strongest instance of a feature, incorporates a certain level of transformation invariance to the image-specific feature.

### 1.6.3   DICTIONARY DESIGN WITH GRAPH EMBEDDING CONSTRAINTS

Though sparse coding can efficiently represent the underlying data, it does not consider the relationship between the training samples. Approaches such as local coordinate coding [39] incorporate the manifold or graph structure of the data indirectly by considering the neighbor relation between the training samples and the dictionary. An alternative approach to representing the relation between the training samples is to construct a neighborhood graph with each sample as a vertex. Several supervised, semi-supervised, and unsupervised machine learning schemes can be unified under the general framework of graph embedding. Incorporating graph embedding principles into sparse coding can provide an improved performance with several learning tasks. Instead of identifying explicit graph-structure preserving projections that reduce the dimensionality, the graph embedding constraints must be imposed on the resulting sparse codes. This general framework can provide highly discriminative sparse codes for a variety of supervised, semi-supervised, and unsupervised graphs.

### 1.6.4   KERNEL SPARSE METHODS

Despite its great applicability, the use of sparse models in complex visual recognition applications presents three main challenges: (a) the linear generative model of sparse coding can be insufficient in describing the non-linear relationship between the image features; (b) in applications such as object recognition, no single descriptor can efficiently model the whole data set; and (c) sparse models require data samples to be represented in the form of feature vectors, and it is not straightforward to extend them to the case of other forms, such as pixel values, matrices, or higher-order tensors. To circumvent these challenges, kernel methods can be incorporated in sparse coding. Kernel methods map the data samples into a high-dimensional feature space, using a non-linear transformation, in which the relationship between the features can be represented using linear models. If the kernel function is chosen such that the resulting feature space is a Hilbert space, computations can be simplified by considering only the similarity between the kernel features. By developing approaches for sparse coding and dictionary learning in the kernel space, novel frameworks can be designed for several computer vision applications.

# CHAPTER 2

# Sparse Representations

Analysis-Synthesis applications in signal and image processing aim to represent data in the most parsimonious terms. In signal analysis, linear models that represent the signals sparsely in a transform domain, such as the Fourier [41] or the wavelet domain [42], have been considered. Hence, sparse representation seeks to approximate an input signal by a linear combination of elementary signals. A common metric considered for measuring the sparsity of the linear combination is to use the number of elementary signals that participate in the approximation. The nature of the signal determines the suitable transform to be applied, such that a sparse representation could be obtained. For example, the wavelet transform has been successfully used for 1-D signals [42] and curvelets have been found to be optimal for representing the edges in an image [43].

In redundant/overcomplete models, the number of basis functions is greater than the dimensionality of the input signals. It has been shown that the approximation power of the model is improved when an overcomplete set of basis functions are used [44],[45]. The other advantages in using redundant representations are that they are well behaved in the presence of noise and they aid in obtaining shift-invariant representations [46].

## 2.1    THE SPARSITY REGULARIZATION

Considering the generative model in (1.1), sparse codes can be obtained either by minimizing the exact $\ell_0$ penalty or its convex surrogate $\ell_1$ penalty as

$$\hat{\mathbf{a}} = \underset{\mathbf{a}}{\operatorname{argmin}} \|\mathbf{a}\|_0 \text{ subj. to } \mathbf{x} = \mathbf{Da}, \tag{2.1}$$

$$\hat{\mathbf{a}} = \underset{\mathbf{a}}{\operatorname{argmin}} \|\mathbf{a}\|_1 \text{ subj. to } \mathbf{x} = \mathbf{Da}, \tag{2.2}$$

where $\|.\|_0$ is the $\ell_0$ norm and $\|.\|_1$ is the $\ell_1$ norm. Since real-world data cannot be expressed exactly using the generative model in (1.1), usually the equality constraints in (2.1) and (2.2) are replaced by using the constraint $\|\mathbf{x} - \mathbf{Da}\|_2^2 \leq \epsilon$, where $\epsilon$ is the error goal of the representation. The exact $\ell_0$ minimization given in (2.1) is a combinatorial problem and that is the major reason why its convex surrogate is often used. In fact, any penalty function from the set $\{\|\mathbf{a}\|_p \mid 0 < p \leq 1\}$ can be shown to be promote sparsity. The shape of the unit $\ell_p$ balls, which are level sets defined by

$$\left( \sum_{i=1}^{K} |a[i]|^p \right)^{1/p} = 1, \tag{2.3}$$

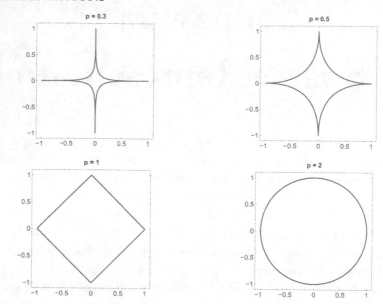

**Figure 2.1:** Unit $\ell_p$ balls for $p = 0.3, 0.5, 1, 2$. Note that the only $\ell_p$ ball that is sparsity promoting and convex is the $\ell_1$ ball.

are shown in Figure 2.1 for various values of $p$. Note that the $p = 0$ cannot be used in (2.3), since the $\ell_0$ norm, that counts the number of non-zero coefficients, is only a pseudonorm. The optimization given in (2.2) can be visualized as the expansion the $\ell_1$ ball until it touches the affine feasible set $\mathbf{x} = \mathbf{Da}$. Considering the various unit balls in Figure 2.1, it can be shown that all points in the $\ell_2$ ball have an equal probability of touching an arbitrary affine feasible set, and, hence, the solution is almost always dense. However, the balls with $p \leq 1$ have a high probability of touching the feasible set at points where most of the coordinates are zero, leading to sparse solutions with high probability. In the rest of this chapter, we restrict our discussion to $\ell_0$ and $\ell_1$ norms.

## 2.1.1    OTHER SPARSITY REGULARIZATIONS

The $\ell_1$ minimization problem in (2.2) can be represented as the penalized optimization

$$\hat{\mathbf{a}} = \underset{\mathbf{a}}{\operatorname{argmin}} \|\mathbf{x} - \mathbf{Da}\|_2^2 + \lambda \|\mathbf{a}\|_1, \tag{2.4}$$

where the parameter $\lambda$ trades off the error and the sparsity of the representation. Suppose there are groups of dictionary atoms that have a high correlation, the $\ell_1$ minimization chooses only one of them and ignores the others. This leads to an unstable representation, as signals that are close in space could have very different representations. Modifying the problem as

$$\hat{\mathbf{a}} = \underset{\mathbf{a}}{\text{argmin}} \ \|\mathbf{x} - \mathbf{D}\mathbf{a}\|_2^2 + \lambda_1 \|\mathbf{a}\|_1 + \lambda_2 \|\mathbf{a}\|^2, \tag{2.5}$$

we have the elastic net regularization that is a combination of the sparsity prior and the ridge-regression penalty [47]. This ridge regression penalty ensures that correlated dictionary atoms are picked together in the solution, leading to a stabler representation. If we have the knowledge that dictionary atoms contribute to the representation in groups, this can be exploited by posing the group lasso problem [48]

$$\hat{\mathbf{a}} = \underset{\mathbf{a}}{\text{argmin}} \ \|\mathbf{x} - \sum_{l=1}^{L} \mathbf{D}_l \mathbf{a}_l\|_2^2 + \lambda \sum_{l=1}^{L} \sqrt{p_l} \|\mathbf{a}_l\|_2, \tag{2.6}$$

where $L$ is the number of groups, $p_l$ depends on the size of the group, $\mathbf{D}_l$ is the $l^{\text{th}}$ dictionary group, and $\mathbf{a}_l$ is the $l^{\text{th}}$ coefficient group. The summation term acts like a sparsity constraint at the group level and all dictionary atoms in a group are either chosen or neglected together. If we further impose sparsity constraints for coefficients in a group, we obtain the sparse group lasso [49]

$$\hat{\mathbf{a}} = \underset{\mathbf{a}}{\text{argmin}} \ \|\mathbf{x} - \sum_{l=1}^{L} \mathbf{D}_l \mathbf{a}_l\|_2^2 + \lambda_1 \sum_{l=1}^{L} \sqrt{p_l} \|\mathbf{a}_l\|_2 + \lambda_2 \|\mathbf{a}\|_1. \tag{2.7}$$

In the sparsity regularizations considered so far, the data samples are considered independent and, hence, the relationship between them is not taken into account. In most cases, real-world data have a strong relationship between each other and utilizing this can be of significant help in stabilizing the representations for inverse problems. Consider the scenario where we have several data samples that are in the same subspace spanned by a group of dictionary atoms, and, hence, have the same sparsity support. This generalization of simple approximation to the case of several input signals is referred to as simultaneous sparse approximation (SSA). Given several input signals, we wish to approximate all the signals at once, using different linear combinations of the same set of elementary signals. The coefficient vectors for sparse coding and SSA are given in Figure 2.2, where it can clearly be seen that SSA prefers to choose the same dictionary for all data samples in the group. For a dictionary $\mathbf{D}$ and the set of input samples given by $\mathbf{X}$, the SSA can be obtained by solving

$$\hat{\mathbf{A}} = \underset{\mathbf{A}}{\text{argmin}} \sum_{i=1}^{k} \|\mathbf{a}^i\|_q^p \ \text{subj. to.} \ \|\mathbf{X} - \mathbf{D}\mathbf{A}\|_F^2 \leq \epsilon, \tag{2.8}$$

where $\mathbf{a}^i$ denotes the $i^{\text{th}}$ row of the coefficient matrix $\mathbf{A}$. The value for the pair $(p, q)$ is chosen as $(1, 2)$ or $(0, \infty)$, and the former leads to a convex norm, whereas the latter actually counts the number of rows with non-zero elements in $\mathbf{A}$.

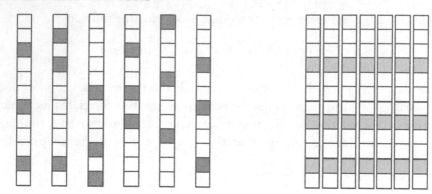

**Figure 2.2:** An example case of coefficient vectors chosen in sparse coding for a group of signals (left) and those of a simultaneous sparse approximation (right). The shaded cells represent non-zero coefficients.

## 2.1.2  NON-NEGATIVE SPARSE REPRESENTATIONS

The representations considered so far had no constraints on the signs of the coefficients. However, there are several applications where we require the model to be *strictly additive*, i.e., the coefficients need to be of the same sign. Some applications of non-negative representations are in image inpainting [50], automatic speech recognition using exemplars [51], protein mass spectrometry [52], astronomical imaging [53], spectroscopy [54], source separation [55], and clustering/semi-supervised learning of data [56, 57], to name a few. The underdetermined system of linear equations with the constraint that the solution is non-negative can be expressed as

$$\mathbf{x} = \mathbf{Gb}, \text{ subj. to } \mathbf{b} \geq 0, \tag{2.9}$$

where $\mathbf{b} \in \mathbb{R}^{K_g}$ is the non-negative coefficient vector and $\mathbf{G} \in \mathbb{R}^{M \times K_g}$ is the dictionary with $K_g > M$. When the solution is sparse, it can be computed using an optimization program similar to (2.1), with an additional non-negativity constraint. The convex program used to solve for this non-negative coefficient vector is given as

$$\min_{\mathbf{b}} \mathbf{1}^T \mathbf{b} \text{ subj. to } \mathbf{x} = \mathbf{Gb}, \alpha \geq \mathbf{0}. \tag{2.10}$$

If the set

$$\{\mathbf{b} | \mathbf{x} = \mathbf{Gb}, \mathbf{b} \geq 0\} \tag{2.11}$$

contains only one solution, any variational function on $\mathbf{b}$ can be used to obtain the solution [58, 59] and $\ell_1$ minimization is not required.

## 2.2    GEOMETRICAL INTERPRETATION

The generative model indicated in (1.1) with sparsity constraints is a non-linear model, because the set of all $S$−sparse vectors is not closed under addition. The sum of two $S$−sparse vectors generally results in a $2S$−sparse vector. Furthermore, sparse models are generalizations of linear subspace models since each sparse pattern represents a subspace, and the union of all patterns represents a union of subspaces. Considering $S$−sparse coefficient vectors obtained from a dictionary of size $M \times K$, the data samples $\mathbf{x}$, obtained using the model (1.1), lie in a union of $\binom{K}{S}$ $S$−dimensional subspaces. In the case of a non-negative sparse model given by (2.9), for $S$−sparse representations, the data samples lie in a union of $\binom{K_g}{S}$ simplical cones. Given a subset $\mathbf{G}_\Omega$ of dictionary atoms, where $\Omega$ is the set of $S$ indices corresponding to the non-zero coefficients, the simplical cone generated by the atoms is given by

$$\left\{ \sum_{j \in \Omega} a[j]\mathbf{g}_j \mid b[j] > 0 \right\}. \tag{2.12}$$

Note that a simplical cone is a subset of the subspace spanned by the same dictionary atoms.

Computing sparse representations using a $\ell_0$ minimization procedure incurs combinatorial complexity, as discussed before, and it is instructive to compare this complexity for the case of general and non-negative representations. For a general representation, we need to identify both the support and the sign pattern, whereas for a non-negative representation, identification of support alone is sufficient. The complexity of identifying the support alone for an $S$−sparse representation is $\binom{K}{S}$, and identification of the sign pattern along with the support incurs a complexity of $\binom{2K-S}{S}$. This is because there are $2K$ signs to choose from and we cannot choose both positive and negative signs for the same coefficient. From figure 2.3 it is clear that the complexity increases as the number of coefficients increases, and the non-negative sparse representation is far less complex compared to the general representation. However, we will see later that the general representation incurs lesser complexity, both with practical convex and greedy optimization procedures. The reason for this is that including additional constraints, such as non-negativity, in an optimization problem usually increases the complexity of the algorithmic procedure in arriving at an optimal solution.

## 2.3    UNIQUENESS OF $\ell_0$ AND ITS EQUIVALENCE TO THE $\ell_1$ SOLUTION

So far, we have discussed, in length, the sparse representations and obtaining representations using $\ell_0$ minimization. However, it is also important to ensure that, for a given dictionary $\mathbf{D}$, such a representation obtained is unique, and both the $\ell_0$ and $\ell_1$ solutions, obtained respectively using (2.1) and (2.2), are equivalent to each other [60, 61].

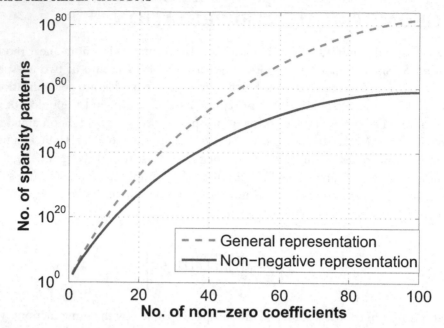

**Figure 2.3:** The number of patterns to be searched for if $\ell_0$ minimization is used to compute sparse representations. Size of the dictionary is $100 \times 200$. Computing non-negative representations always incurs lesser complexity compared to general representations.

To analyze the uniqueness of the solution for an arbitrary (in this case overcomplete) dictionary $\mathbf{D}$, assume that there are two suitable representations for the input signal $\mathbf{x}$,

$$\exists \mathbf{a}_1 \neq \mathbf{a}_2 \quad \text{such that} \quad \mathbf{x} = \mathbf{D}\mathbf{a}_1 = \mathbf{D}\mathbf{a}_2. \tag{2.13}$$

Hence, the difference of the representations $\mathbf{a}_1 - \mathbf{a}_2$ must be in the null space of the representation $\mathbf{D}(\mathbf{a}_1 - \mathbf{a}_2) = 0$. This implies that some group of elements in the dictionary should be linearly dependent. To quantify the relation, we define the *spark* of a matrix. Given a matrix, the spark is defined as the smallest number of columns that are linearly dependent. This is quite different from the *rank* of a matrix, which is the largest number of columns that are linearly independent. If a signal has two different representations, as in (2.13), we must have $\|\mathbf{a}_1\|_0 + \|\mathbf{a}_2\|_0 \geq \text{spark}(\mathbf{D})$. From this argument, if any representation exists satisfying the relation $\|\mathbf{a}_1\|_0 < \text{spark}(\mathbf{D})/2$, then, for any other representation $\mathbf{a}_2$, we have $\|\mathbf{a}_2\|_0 > \text{spark}(\mathbf{D})/2$. This indicates that the sparsest representation is $\mathbf{a}_1$. To conclude, a representation is the sparsest possible if $\|\mathbf{a}\|_0 < \text{spark}(\mathbf{D})/2$. Assuming that the dictionary atoms are normalized to unit $\ell_2$ norm, let us define the Gram matrix for the dictionary as $\mathbf{H} = \mathbf{D}^T\mathbf{D}$ and denote the coherence as the maximum magnitude of

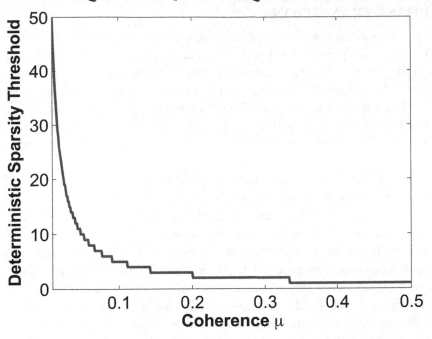

**Figure 2.4:** Deterministic sparsity threshold with respect to the coherence of the dictionary.

the off-diagonal elements

$$\mu = \max_{i \neq j} |h_{i,j}|. \tag{2.14}$$

Then we have the bound spark$(\mathbf{D}) > 1/M$ [60], and, hence, it can be inferred that the representation obtained from $\ell_0$ minimization is unique and equivalent to $\ell_1$ minimization if

$$\|\mathbf{a}\|_0 \leq \frac{1}{2}\left(1 + \frac{1}{\mu}\right). \tag{2.15}$$

This is referred to as the *deterministic sparsity threshold*, since it holds true for all sparsity patterns and non-zero values in the coefficient vectors. This threshold is illustrated in Figure 2.4 for various values of $\mu$. Since the general representation also encompasses the non-negative case, the same bound holds true for non-negative sparse representations. Furthermore, the threshold is the same for $\ell_1$ minimization, as well as greedy recovery algorithms, such as the Orthogonal Matching Pursuit (OMP). The deterministic sparsity threshold scales at best as $\sqrt{M}$ with increasing values of $M$. Probabilistic or Robust sparsity thresholds, on the other hand, scale in the order of $M/\log K$ [62] and break the square-root bottleneck. However, the trade-off is that the unique recovery using $\ell_1$ minimization is only assured with high probability, and robust sparsity thresholds for unique recovery using greedy algorithms, such as OMP, are also unknown.

### 2.3.1   PHASE TRANSITIONS

Deterministic thresholds are too pessimistic, and, in reality, the performance of sparse recovery is much better than that predicted by the theory. Robust sparsity thresholds are better, but still restrictive, as they are not available for several greedy sparse recovery algorithms. Phase-transition diagrams describe sparsity thresholds at which the recovery algorithm transitions from a high probability of success to a high probability of failure, for various values of the ratio $M/K$ (undersampling factor) ranging from 0 to 1. For random dictionaries and coefficient vectors whose entries are realized from various probability distributions, empirical phase transitions can be computed by finding the points at which the fraction of success for sparse recovery is 0.5 with respect to a finely spaced grid of sparsity and undersampling factors [63].

Asymptotic phase transitions can be computed based on the theory of polytopes [58, 63, 64, 65]. For $K \to \infty$, when the dictionary entries are derived from $\mathcal{N}(0, 1)$ and the non-zero coefficients are signs ($\pm 1$), asymptotic phase transitions are shown in Figure 2.5 for $\ell_1$ minimization algorithms. It can easily be shown that the unconstrained $\ell_1$ minimization corresponds to the cross-polytope object, and non-negative $\ell_1$ minimization corresponds to the simplex object. Clearly, imposing non-negativity constraint gives an improved phase transition when compared to the unconstrained case. Furthermore, empirical phase transitions computed for Rademacher, partial Hadamard, Bernoulli, and random Ternary ensembles are similar to those computed for Gaussian i.i.d. ensembles [63]. The phase transitions of several greedy recovery algorithms have been analyzed and presented in [66], and optimal tuning of sparse approximation algorithms that use iterative thresholding have been performed by studying their phase-transition characteristics [67].

## 2.4   NUMERICAL METHODS FOR SPARSE CODING

When $\ell_0$ norm is used as the cost function, exact determination of the sparsest representation is an NP-hard problem [68], and the complexity of search becomes intractable, even for a moderate number of non-zero coefficients, as evident from Figure 2.3. Hence, a number of numerical algorithms that use $\ell_1$ approximation and greedy procedures have been developed to solve these problems. Note that most of these algorithms have a non-negative counterpart and it is straightforward to develop them by appropriately placing the non-negativity constraint.

Some of the widely used methods for computing sparse representations include the Matching Pursuit (MP) [69], Orthogonal Matching Pursuit (OMP) [70], Basis Pursuit (BP) [71], FOCUSS [72], Feature Sign Search (FSS) [73], Least Angle Regression (LARS) [74], and iterated shrinkage algorithms [75, 76]. Before describing the sparse coding algorithms, we will present an overview of the optimality conditions used by these procedures.

The recovery performance of $\ell_1$ minimization (BP) and greedy (OMP) methods are compared in Figure 2.6 for general and non-negative representations. It can be seen that non-negative representations result in a better performance, in terms of recovery, when compared to general representations, both with BP and OMP.

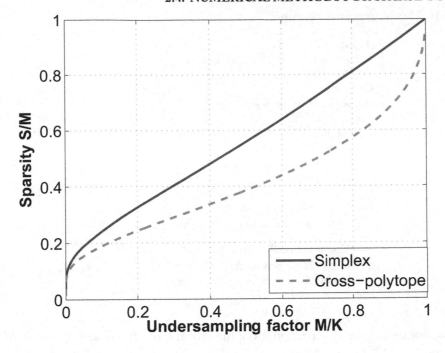

**Figure 2.5:** Asymptotic phase transitions for non-negative (simplex) and general $\ell_1$ minimization (cross-polytope) when the dictionary elements are derived from i.i.d. Gaussian $\mathcal{N}(0, 1)$ and non-zero coefficients are signs.

## 2.4.1 OPTIMALITY CONDITIONS

Considering the $\ell_0$ optimization problem in (2.1), it can be shown that a unique and, hence, an optimal solution can be obtained using BP, MP, and OMP algorithms [70, 77] if the condition given in (2.15) is satisfied. For algorithms that use the penalized $\ell_1$ formulation given in (2.4), the optimality condition is obtained computing the following sub-gradient set:

$$2\mathbf{D}^T (\mathbf{Da} - \mathbf{x}) + \lambda\mathbf{p}, \tag{2.16}$$

and ensuring that it contains the zero vector. Here $\mathbf{p}$ is the sub-gradient of $\|\mathbf{a}\|_1$ and is defined as

$$p[i] = \begin{cases} +1 & a[i] > 0 \\ [-1, +1] & a[i] = 0. \\ -1 & a[i] < 0 \end{cases} \tag{2.17}$$

Hence, the optimality conditions can be simplified as [78]

$$2\mathbf{d}_i^T (\mathbf{x} - \mathbf{Da}) = \lambda\text{sign}(a[i]), \text{ if } a[i] \neq 0, \tag{2.18}$$

$$2|\mathbf{d}_i^T (\mathbf{x} - \mathbf{Da})| < \lambda, \text{ otherwise.} \tag{2.19}$$

**Table 2.1:** Matching pursuit algorithm

**Input:**
Input signal, $\mathbf{x} \in \mathbb{R}^M$, Dictionary, $\mathbf{D}$.

**Output:**
Coefficient vector, $\mathbf{a} \in \mathbb{R}^K$.

**Initialization:**
Initial residual: $\mathbf{r}^{(0)} = \mathbf{x}$.
Initial coefficient vector: $\mathbf{a} = \mathbf{0}$.
Loop index: $l = 1$.

**while** convergence not reached
 —Determine an index $k_l$: $k_l = \underset{k}{\operatorname{argmax}} |\langle \mathbf{r}^{(l-1)}, \mathbf{d}_k \rangle|$.
 —Update the coefficient: $a[k_l] \leftarrow a[k_l] + \langle \mathbf{r}^{(l-1)}, \mathbf{d}_{k_l} \rangle$.
 —Compute the new residual: $\mathbf{r}^{(l)} = \mathbf{r}^{(l-1)} - \langle \mathbf{r}^{(l-1)}, \mathbf{d}_{k_l} \rangle \mathbf{d}_{k_l}$.
 —Update the loop counter: $l = l + 1$.
**end**

These conditions are used as criterion for optimality and coefficient selection by LARS and FSS algorithms.

### 2.4.2  BASIS PURSUIT

Basis Pursuit (BP) is a linear programming approach [79] that solves (2.2) to find the minimum $\ell_1$ norm representation of the signal [71].

A standard linear program (LP) is a constrained optimization problem with affine objective and constraints. We can convert the problem in (2.2) to a standard LP, by adding an auxiliary variable $\mathbf{u} \in \mathbb{R}^K$ as

$$\min_{\mathbf{u}} \mathbf{1}^T \mathbf{u} \quad \text{subject to} \quad \mathbf{D}\mathbf{a} = \mathbf{x}, \ \mathbf{u} \geq 0, \ -\mathbf{u} \leq \mathbf{a} \leq \mathbf{u}. \tag{2.20}$$

Relating (2.2) to this LP, the problem is to identify which elements in $\mathbf{a}$ should be zero. To solve this, both *simplex* methods and *interior point* methods have been used [71]. Geometrically, the collection of feasible points is a convex polyhedron or a *simplex*. The simplex method works by moving around the boundary of the simplex, jumping from one vertex of the polyhedron to another, where the objective is better. On the contrary, interior point methods start from the interior of the simplex. Since, the solution of LP is at an extreme point of the simplex, as the algorithm converges, the estimate moves toward the boundary. When the representation is not exact and the error goal $\epsilon$ is known, the sparse code can be computed by solving the quadratically constrained problem

$$\hat{\mathbf{a}} = \underset{\mathbf{a}}{\operatorname{argmin}} \|\mathbf{a}\|_1 \ \text{subj. to} \ \|\mathbf{x} - \mathbf{D}\mathbf{a}\|_2 \leq \epsilon. \tag{2.21}$$

This is referred to as Basis Pursuit Denoising (BPDN) and the solution to this can be obtained using several efficient procedures [71]. For a data vector $\mathbf{x}$ of dimensionality $M$, corrupted with additive white Gaussian noise (AWGN) of variance $\sigma^2$, the squared error goal $\epsilon^2$ is fixed at $(1.15\sigma^2)M$. Using this error goal will ensure that each component in the AWGN noise vector will lie within the range $[-1.15\sigma, 1.15\sigma]$ with a probability of 0.75. Hence, we will have a very low chance of picking noise as a part of the representation. For the non-negative case, the problem given in (2.10) can be directly solved as an LP and the BPDN problem also can be modified easily in order to incorporate non-negativity.

### 2.4.3  GREEDY PURSUIT METHODS

Greedy procedures for computing sparse representations operate by choosing the atom that is most strongly correlated to the current target vector, removing its contribution and iterating. Hence, they make a sequence of locally optimal choices in an effort to obtain a global optimal solution. There are several versions of such greedy algorithms, some that are aggressive and remove all of the contributions of the particular atom from the target, and some are less aggressive and remove only a part of the contribution. For orthonormal dictionaries, even the most aggressive greedy algorithms perform well, whereas, for overcomplete dictionaries, those that are more careful in fixing the coefficient result in a better approximation. The greedy methods for solving sparse approximation problems generalize this simple idea to the case of any arbitrary dictionary. A clear advantage with greedy algorithms is that there is a flexibility to fix the number of non-zero coefficients, and/or the error goal $\epsilon$, which is not the case with BP or BPDN.

**Matching Pursuit (MP)**

This is the simplest of greedy pursuit algorithms and was proposed by Mallat and Zhang [69]. The steps involved in the MP algorithm are shown in Table 2.1. The algorithm begins by setting the initial residual to the input signal $\mathbf{x}$ and making a trivial initial approximation. It then iteratively chooses the best correlating atom to the residual and updates the residual correspondingly. In every iteration, the algorithm takes into account only the current residual and not the previously selected atoms, thereby making this step greedy. It is important to note that MP might select the same dictionary atom many times, when the dictionary is not orthogonal. The contribution of an atom to the approximation is the inner product itself, as in the case of an orthogonal expansion. The residual is updated by removing the contribution of the component in the direction of the atom $\mathbf{d}_{k_l}$. At each step of the algorithm a new approximant of the target signal, $\mathbf{x}^{(l)}$, is calculated based on the relationship

$$\mathbf{x}^{(l)} = \mathbf{x} - \mathbf{r}^{(l)}. \tag{2.22}$$

When the dictionary is orthonormal, the representation $\mathbf{x}^{(S)}$ is always an optimal and unique $S$-term representation of the signal [70]. It has been shown that, for general dictionaries, the norm of the residual converges to zero [80].

**Table 2.2:** Orthogonal matching pursuit algorithm

---

**Input:**
  Input signal, $\mathbf{x} \in \mathbb{R}^N$, Dictionary, $\mathbf{D}$.

**Output:**
  Coefficient vector, $\mathbf{a} \in \mathbb{R}^K$.

**Initialization:**
  Initial residual, $\mathbf{r}^{(0)} = \mathbf{x}$.
  Index set, $\Omega = \{\}$.
  Loop index, $l = 1$.

**while** convergence not reached
  —Determine an index $k_l$: $k_l = \text{argmax}_k |\langle \mathbf{r}^{(l-1)}, \mathbf{d}_k \rangle|$.
  —Update the index set: $\Omega \leftarrow \Omega \cup k_l$.
  —Compute the coefficient: $\mathbf{a} = \text{argmin}_{\mathbf{a}} \left\| \mathbf{x} - \sum_{j=1}^l a[k_j]\mathbf{d}_{k_j} \right\|_2$.
  —Compute the new residual: $\mathbf{r}^{(l)} = \mathbf{x} - \sum_{j=1}^l a[k_j]\mathbf{d}_{k_j}$.
  —Update the loop counter: $l = l + 1$.
**end**

---

**Orthogonal Matching Pursuit (OMP)**

This algorithm introduces a least squares minimization to each step of the MP algorithm, in order to ensure that the best approximation is obtained over the atoms that have already been chosen [81, 82, 83]. This is also referred to as forward selection algorithm [84] and the steps involved in this algorithm are provided in Table 2.2. The initialization of OMP is similar to that of MP, with a difference that an index set is initialized to hold the indices of atoms chosen at every step. The atom selection procedure of this algorithm is also greedy, as in MP. The index set is updated by adding the index of the currently chosen atom to the list of atoms that have already been chosen. Unlike the MP algorithm, where the correlations themselves were the coefficients, OMP computes the coefficients by solving a least squares problem. The most important behavior of OMP is that the greedy selection always picks an atom that is linearly independent from the atoms already chosen. In other words,

$$\langle \mathbf{r}^{(l)}, \mathbf{d}_{k_j} \rangle = 0 \quad \text{for } j = 1, \dots, l. \tag{2.23}$$

As a result, the residual must be equal to zero in $N$ steps. Further, the atoms corresponding to index set $\Omega$ result in a full rank matrix and, hence, the least squares solution is unique. Since the residual is orthogonal to the atoms already chosen, OMP ensures that an atom is not chosen more than once in the approximation. The solution of the least squares problem determines the approximant,

$$\mathbf{x}^{(l)} = \sum_{j=1}^{l} a[k_j] \mathbf{d}_{k_l}. \tag{2.24}$$

A non-negative version of the OMP algorithm has been proposed in [85] as a greedy procedure for recovering non-negative sparse representations. Several pursuit methods have been recently proposed to improve upon the performance of OMP. One such procedure, the compressive sampling matching pursuit (CoSaMP) [86], follows a procedure similar to the OMP, but selects multiple columns of the dictionary and prunes the set of active columns in each step.

*Simultaneous Orthogonal Matching Pursuit (S-OMP)*

Using a set of $T$ observations from a single source, we build the signal matrix $\mathbf{X} = [\mathbf{x}_1 \mathbf{x}_2 \ldots \mathbf{x}_T]$. The S-OMP algorithm identifies a representation such that all the signals in $\mathbf{X}$ use the same set of dictionary atoms from $\mathbf{D}$ with different coefficients, minimizing the error given by $\|\mathbf{X} - \mathbf{D}\mathbf{A}\|_F^2$. Here, $\mathbf{A} \in \mathbb{R}^{K \times T}$ indicates the coefficient matrix, where each column corresponds to the coefficients for a single observation and $\|.\|_F$ refers to the Frobenius norm. This algorithm reduces to a simple OMP, when $T = 1$. The atom selection step of this algorithm is also greedy, similar to the cases of the MP and the OMP algorithms discussed earlier. Each atom is chosen such that it has the maximum sum of absolute correlations with the current target signal matrix. The basic intuition behind this is that the chosen atom will capture a lot of energy from each column of the current target matrix [87]. Hence, this approach will be very effective, if the same set of atoms is used to approximate all the columns of $\mathbf{X}$.

## Least Angle Regression (LARS)

The LARS procedure computes the solution to penalized $\ell_1$ optimization in (2.4) by selecting the active coefficients one at a time and performing a least squares procedure at every step to re-estimate them. This method is closely related to the OMP or the forward selection algorithm, but it is less greedy in reducing the residual error. Another closely related method to OMP and LARS is the forward stagewise algorithm, which takes thousands of tiny steps as it moves toward the solution. All of these algorithms update their current approximant of the signal by taking a step in the direction of the largest correlation of the dictionary atom with the current approximant

$$\mathbf{p}^{(l+1)} = \mathbf{p}^{(l)} + \tau \cdot \text{sign}(\langle \mathbf{r}^{(l)}, \mathbf{d}_{k_l} \rangle), \text{ where } l = \underset{j}{\text{argmax}} \, |\langle \mathbf{r}^{(j)}, \mathbf{d}_{k_j} \rangle|. \tag{2.25}$$

For MP, the step size $\tau$ is the same as $|\langle \mathbf{r}^{(l)}, \mathbf{d}_{k_l} \rangle|$, whereas for OMP the current approximant is re-estimated using a least squares method. The forward stagewise procedure is overly careful and fixes $\tau$ as a small fraction of the MP step size. LARS is a method that strikes a balance in step-size selection.

The idea behind the LARS algorithm can be described as follows. The coefficients are initialized as zero and the dictionary atom $\mathbf{d}_{k_1}$ that is most correlated with the signal $\mathbf{x}$ is chosen.

The largest step possible is taken in the direction of $\mathbf{d}_{k_1}$ until another dictionary atom $\mathbf{d}_{k_2}$ has as much correlation as $\mathbf{d}_{k_1}$ with the current residual. Then the algorithm proceeds in the equiangular direction between $\mathbf{d}_{k_1}$ and $\mathbf{d}_{k_2}$ until a new dictionary atom $\mathbf{d}_{k_3}$ has as much correlation with the residual as the other two. Therefore, at any step in the process, LARS proceeds equiangularly between the currently chosen atoms. The difference between MP and LARS is that, in the former case, a step is taken in the direction of the atom with maximum correlation with the residual, whereas, in the latter case, the step taken is in the direction that is equiangular to all the atoms that are most correlated with the residual. Note that the LARS algorithm always adds coefficients to the active set and never removes any, and, hence, it can only produce a sub-optimal solution for (2.4). Considering the optimality constraints in (2.18) and (2.19), the LARS algorithm implicitly checks for the all the conditions except the sign condition in (2.18). Hence, in order to make the LARS solution the same as the optimal solution for (2.4), a small modification that provides a way for the deletion of coefficients from the current non-zero coefficient set has to be included.

The LARS algorithm can produce a whole set of solutions for (2.4) for $\lambda$ varying between 0, when $M$ dictionary atoms are chosen and a least squares solution is obtained, to its original value, where a sparse solution is obtained. If the computations are arranged properly, LARS is a computationally cheap procedure with complexity of the same order as that of least squares.

### 2.4.4   FEATURE-SIGN SEARCH

Feature-sign search is an efficient sparse coding method that solves the penalized $\ell_1$ optimization in (2.4) by maintaining an *active set* of non-zero coefficients and corresponding signs, and searches for the optimal active set and coefficient signs. In each *feature-sign step*, given an active set, an analytical solution is computed for the resulting quadratic program. Then, the active set and the signs are updated using a discrete line search and by selecting coefficients that promote optimality of the objective. In this algorithm, the sparse code $\mathbf{a}$ is initialized to the zero vector and the active set $\mathcal{A}$ is initialized to an empty set. The sign vector is denoted by $\boldsymbol{\theta}$ and its entries are defined as $\theta[i] = \text{sign}(a[i])$, which is one of the elements from $\{-1, 0, +1\}$. The algorithm consists of the following steps.

*Step 1:* An element of the non-active set that results in the maximum change in the error term is chosen as

$$i = \underset{i}{\operatorname{argmax}} \left| \frac{\partial \|\mathbf{x} - \mathbf{Da}\|_2}{\partial a[i]} \right|, \tag{2.26}$$

and add $i$ to the active set if it improves the objective locally. This is obtained when the subgradient given by (2.16) is less than 0 for a given index $i$. Hence, we append $i$ to $\mathcal{A}$, if $\frac{\partial \|\mathbf{x} - \mathbf{Da}\|}{\partial a[i]} > \lambda/2$ setting $\theta[i] = -1$, or if $\frac{\partial \|\mathbf{x} - \mathbf{Da}\|}{\partial a[i]} < -\lambda/2$, setting $\theta[i] = 1$.

*Step 2:* Using coefficients only from the active set, let $\hat{\mathbf{D}}$ be the sub-dictionary corresponding to the active set, and let $\hat{\mathbf{a}}$, $\hat{\boldsymbol{\theta}}$ be the active coefficient and sign vectors. The strategy used in this feature-sign step is to find the new set of coefficients that have the sign pattern consistent with

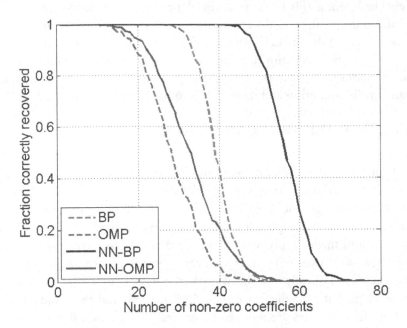

**Figure 2.6:** Recovery performance for general and non-negative sparse representations. The dictionary is obtained as realizations from $\mathcal{N}(0, 1)$ and is of size $100 \times 200$. Non-zero coefficients are realized from a uniform random distribution. The performance is averaged over 100 iterations for each sparsity level.

the sign vector. In order to do this, we first solve the unconstrained optimization

$$\min_{\hat{\mathbf{a}}} \|\mathbf{x} - \hat{\mathbf{D}}\hat{\mathbf{a}}\|_2^2 + \frac{\lambda}{2}\hat{\theta}^T \hat{\mathbf{a}} \tag{2.27}$$

and obtain $\hat{\mathbf{a}}_{new}$. Denoting the objective in (2.27) as $f(\hat{\mathbf{a}}, \hat{\theta})$, it is easy to see that $f(\hat{\mathbf{a}}_{new}, \hat{\theta}) < f(\hat{\mathbf{a}}, \hat{\theta})$, since $\hat{\mathbf{a}}_{new}$ is the optimal solution of (2.27). If $\text{sign}(\hat{\mathbf{a}}_{new}) = \hat{\theta}$, $\hat{\mathbf{a}}_{new}$ is the updated solution. Else, a line search will be performed between $\hat{\mathbf{a}}$ and $\hat{\mathbf{a}}_{new}$, and the points where at least one coefficient in the vector becomes zero are noted. Among these, the one with the lowest objective value, $\hat{\mathbf{a}}_l$, is chosen as the new coefficient vector. Because of the convexity of $f$, we have $f(\hat{\mathbf{a}}_l, \hat{\theta}) < f(\hat{\mathbf{a}}, \hat{\theta})$, and, hence, the objective value decreases. In this case, elements are removed from the active set since at least one of the coefficients in the active set has become zero, and $\theta = \text{sign}(\mathbf{a})$.

*Step 3:* Optimality conditions need to be checked for both non-zero and zero coefficients. For non-zero coefficients belonging to the set $\mathcal{A}$, the condition is given by (2.18). If this does not hold true, *Step 2* is repeated without any new activation. For zero coefficients, the condition given by (2.19) is checked and, if this is not true, the algorithm proceeds to *Step 1*. However, if both optimality conditions are satisfied, the algorithm exits with the current solution as optimal.

It can be seen that the feature-sign step (*Step 2*) of the algorithm strictly reduces the objective, and *Step 3* of the algorithm mandates that the iterative procedure stops only when an optimal solution is attained. It can be shown, using a simple proof, that the FSS algorithm produces a globally optimal solution and more details on this can be found in [73].

## 2.4.5  ITERATED SHRINKAGE METHODS

The iterative optimization algorithms based on gradient descent and interior point methods are computationally intensive for higher dimensional signals. An alternative approach to solving the sparse approximation efficiently extends the wavelet shrinkage method proposed in [88]. Initially developed to perform signal denoising, the shrinkage procedure begins by computing a forward transform (unitary) of the noisy input signal. This is followed by performing a shrinkage operation on the coefficients and finally applying the inverse transform to reconstruct the signal. Though orthogonal matrices were used initially, it has been found that shrinkage works well, even with redundant dictionaries. Typically, the shrinkage function is a curve that maps an input value to an output value, such that the values near the origin are zero and those outside the interval are shrunk. When unitary matrices are replaced by general dictionaries, it has been found that the shrinkage operations need to be performed iteratively. Several methods for iterated shrinkage have been proposed in the literature [89]: (a) Proximal point method with surrogate functions; (b) expectation maximization approaches; (c) iterative-reweighted least-squares-based method; (d) parallel coordinate descent algorithm; and (e) stagewise OMP.

# CHAPTER 3

# Dictionary Learning: Theory and Algorithms

Dictionaries used for sparse representations can be constructed based on the mathematical model that generates the data. There are also methods available to tune the parameters of a pre-defined dictionary, such that the performance over the given set of data is optimized. However, dictionaries that are learned directly from the data result in an improved performance compared to both pre-defined as well as tuned dictionaries. This chapter will focus exclusively on learned dictionaries and their applications in various image processing tasks.

Pre-defined dictionaries can be constructed using bases from transforms such as the discrete cosine transform (DCT), wavelet, and curvelet [90] and they have been used successfully in various image reconstruction applications. In addition, predefined dictionaries can also be constituted as a union of multiple bases, where each basis represents certain features in the image well [91, 92]—this generally results in improved performances compared to using a single basis. A well-known example for tunable dictionaries is wavelet packets, which generate a library of bases for a given wavelet function, and the basis that results in the optimal performance for the data can be chosen [93]. In their celebrated work, Olshausen and Field proposed a framework to learn an overcomplete set of individual vectors optimized to the training data for sparse coding [13]. Since then, a wide range of dictionary learning algorithms for sparse coding have been proposed in the literature, some of which are tailored for specific applications. A few important algorithms for learned dictionaries that are useful in general signal and image processing applications are discussed in this chapter. Chapter 5 will deal with dictionary learning procedures that are useful in visual recognition tasks.

Assume that the training data $\mathbf{x}$ is obtained from a probability space obeying the generative model posed in (1.1), the dictionary learning problem can be expressed as minimizing the objective

$$g(\mathbf{D}) = \mathbf{E}_{\mathbf{x}}[h(\mathbf{x}, \mathbf{D})], \tag{3.1}$$

where the columns of $\mathbf{D}$, referred to as dictionary atoms, are constrained as $\|\mathbf{d}_j\|_2 \leq 1, \forall j$. The cost of the penalized $\ell_1$ minimization given in (2.4) is denoted by $h(\mathbf{x}, \mathbf{D})$. If the continuous probability distribution is unknown and we only have $T$ training samples $\{\mathbf{x}_i\}_{i=1}^{T}$, equiprobable

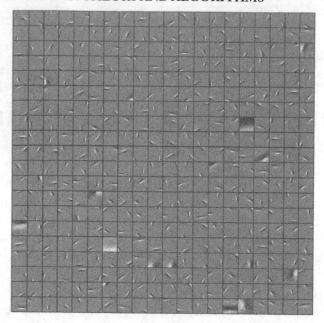

**Figure 3.1:** An example dictionary learned using the algorithm proposed by Olshausen and Field [94].

with a mass of $\frac{1}{T}$, (3.1) can be modified as the empirical cost function,

$$\hat{g}(\mathbf{D}) = \frac{1}{T} \sum_{i=1}^{T} h(\mathbf{x}_i, \mathbf{D}).$$ (3.2)

Typically, dictionary learning algorithms solve for the sparse codes and obtain the dictionary by minimizing $\hat{g}(\mathbf{D})$, repeating the steps until convergence [95, 96, 97].

Given the sparse codes, dictionary learning can be posed as the convex problem

$$\min_{\mathbf{D}} \sum_{i=1}^{T} \|\mathbf{x}_i - \mathbf{D}\mathbf{a}_i\|_2^2 \text{ subj. to } \|\mathbf{d}_j\|_2 \leq 1, \forall j,$$ (3.3)

where $\mathbf{D} = [\mathbf{d}_1 \ \mathbf{d}_2 \ \dots \ \mathbf{d}_K]$ is the dictionary matrix and $S$ is the sparsity of the coefficient vector. Denoting $\mathbf{X} = [\mathbf{x}_1 \ \mathbf{x}_2 \ \dots \ \mathbf{x}_T]$ as the collection of $T$ training vectors and $\mathbf{A} = [\mathbf{a}_1 \ \mathbf{a}_2 \ \dots \ \mathbf{a}_T]$ as the coefficient matrix, the objective in (3.3) can be re-written as $\|\mathbf{X} - \mathbf{D}\mathbf{A}\|_F^2$, where $\|.\|_F$ denotes the Frobenius norm. Learned dictionaries have been successfully applied in image deblurring, compression, denoising, inpainting, and super-resolution [20].

# 3.1   DICTIONARY LEARNING AND CLUSTERING

In the clustering problem, the goal is to partition the data into a desired number of disjoint subsets (referred as clusters), such that data within each cluster are similar to each other, where the degree of similarity is measured based on the distortion function under consideration. The centroid for each cluster represents a prototype geometric structure and each data vector is assigned to the nearest cluster center based on the distortion measure. In contrast, during sparse approximation, each data vector can be assigned to more than one cluster center (dictionary atom) and the number of dictionary atoms that participate in the representation also depends on the penalty in the problem formulation. The dictionary learning and sparse coding optimization posed in (3.2) is a generalization of clustering.

## 3.1.1   CLUSTERING PROCEDURES

K-means algorithm, a commonly used clustering procedure, partitions data into different clusters such that the sum of squares of Euclidean distances between the data and their corresponding cluster centers are minimized. Given the data sample $\mathbf{x}_i$ and the cluster center $\mathbf{d}_j$, the distortion function for K-means is given by $\|\mathbf{x}_i - \mathbf{d}_j\|_2^2$. Comparing this with (3.2), it can be seen that K-means can be obtained if we assume that each coefficient vector is 1-sparse, the non-zero coefficient takes the value 1, and the norm of the dictionary atoms is unconstrained. Hence, K-means represents each data sample using a representative mean value. If the coefficient value is left unconstrained and the cluster center is constrained to be of unit $\ell_2$ norm, we obtain the K-lines clustering algorithm [98]. This procedure attempts to fit $K$ $1-$dimensional linear subspaces, represented by the normalized cluster centers $\{\mathbf{d}_j\}_{j=1}^K$, to the data. The distortion between the cluster center $\mathbf{d}_j$ and the data sample $\mathbf{x}_i$ is given by the measure

$$\rho(\mathbf{x}_i, \mathbf{d}_j) = \|\mathbf{x}_i - \mathbf{d}_j(\mathbf{d}_j^T \mathbf{x}_i)\|_2^2. \tag{3.4}$$

This algorithm computes the cluster centers and the memberships of data samples to cluster centers such that the sum of distortions between the data and their corresponding cluster centers is minimized.

The K-lines algorithm also proceeds in two steps after initialization—the cluster assignment and the cluster centroid update. The index of the cluster center corresponding to the data sample $\mathbf{x}_i$ is obtained using the function $\mathcal{H}(\mathbf{x}_i) = \text{argmin}_j \rho(\mathbf{x}_i, \mathbf{d}_j)$, which is equivalent to $\mathcal{H}(\mathbf{x}_i) = \text{argmax}_j |\mathbf{x}_i^T \mathbf{d}_j|$. We also define the membership set $\mathcal{C}_j = \{i | \mathcal{H}(\mathbf{y}_i) = j\}$ containing training vector indices corresponding to the cluster $j$. The update of a cluster centroid can be expressed as

$$\mathbf{d}_j = \underset{\mathbf{d}_j}{\text{argmin}} \sum_{i \in \mathcal{C}_j} \|\mathbf{x}_i - \mathbf{d}_j(\mathbf{x}_i^T \mathbf{d}_j)\|_2^2. \tag{3.5}$$

As a solution to (3.5), the centroid of a given cluster is computed as the left singular vector corresponding to the largest singular value of the data matrix $\mathbf{Y}_j = [\mathbf{y}_i]_{i \in \mathcal{C}_j}$. Later on, we will see in Chapter 5 that this expensive singular value decomposition (SVD) procedure can be substituted using iterative linear updates and, in fact, such an approximation is necessary for certain formulations.

## 3.1.2   PROBABILISTIC FORMULATION

The general problem of dictionary learning from a set of training data can be formulated as an Expectation-Maximization (EM) procedure [99]. The probabilistic source model for the EM formulation is given by the generative model, $\mathbf{x} = \mathbf{D}\mathbf{a} + \mathbf{n}$, and the $T$ independent realizations from the random vector $\tilde{\mathbf{x}}$ are given by the matrix of data samples, $\mathbf{X}$. Similar to our assumption in the beginning of this chapter, here also, we can assume that the $T$ realizations are equiprobable. The noise vector $\mathbf{n} \sim \mathcal{N}(0, \sigma^2 \mathbf{I}_M)$, where $\sigma^2$ is the known noise variance and $\mathbf{I}_M$ is an identity matrix of size $M \times M$. In order to enforce sparsity, an appropriate prior is placed on each coefficient vector $\mathbf{a}_i$. The only parameter to be estimated by the EM algorithm is $\mathbf{D}$ and, hence, the parameter set $\Theta = \{\mathbf{D}\}$. The likelihood for $\mathbf{x}_i$ at the iteration $l$ of the EM algorithm is given by

$$p(\mathbf{x}_i | \mathbf{a}_i, \Theta^{(l)}) = \mathcal{N}(\mathbf{D}^{(l)} \mathbf{a}_i, \sigma^2 \mathbf{I}_M). \tag{3.6}$$

The posterior probability of the coefficient vector for the current iteration is computed as

$$p\left(\mathbf{a}_i | \mathbf{x}_i, \Theta^{(l)}\right) = \frac{p\left(\mathbf{a}_i, \mathbf{x}_i | \Theta^{(l)}\right)}{p\left(\mathbf{x}_i | \Theta^{(l)}\right)} = \frac{p\left(\mathbf{a}_i\right) p\left(\mathbf{x}_i | \mathbf{a}_i, \Theta^{(l)}\right)}{\int_{\mathbf{a}_i} p\left(\mathbf{x}_i, \mathbf{a}_i | \Theta^{(l)}\right) d\mathbf{a}_i}. \tag{3.7}$$

The expectation that needs to be maximized to find the parameters $\Theta$ for the next iteration $l + 1$ is given by

$$Q(\Theta | \Theta^{(l)}) = \mathbf{E}_{p(\{\mathbf{a}_i\}_{i=1}^T | \{\mathbf{x}_i\}_{i=1}^T, \Theta^{(l)})} \left[ \log p\left(\{\mathbf{x}_i\}_{i=1}^T, \{\mathbf{a}_i\}_{i=1}^T | \Theta\right) \right]. \tag{3.8}$$

Since the data vectors and the coefficients are independent and equiprobable we can express

$$\log p\left(\{\mathbf{x}_i\}_{i=1}^T, \{\mathbf{a}_i\}_{i=1}^T | \Theta\right) = \sum_{i=1}^T \log p\left(\mathbf{x}_i, \mathbf{a}_i | \Theta\right), \tag{3.9}$$

$$p\left(\{\mathbf{a}_i\}_{i=1}^T | \{\mathbf{x}_i\}_{i=1}^T, \Theta^{(l)}\right) = \prod_{i=1}^T p\left(\mathbf{a}_i | \mathbf{x}_i, \Theta^{(l)}\right). \tag{3.10}$$

From (3.8), (3.9) and (3.10) we have

$$Q(\Theta | \Theta^{(l)}) = \sum_{i=1}^T \int_{\mathbf{a}_i} \log p\left(\mathbf{x}_i, \mathbf{a}_i | \Theta\right) p\left(\mathbf{a}_i | \mathbf{x}_i, \Theta^{(l)}\right) d\mathbf{a}_i,$$

$$= \sum_{i=1}^T \int_{\mathbf{a}_i} \left[ \log \mathcal{N}\left(\mathbf{x}_i | \mathbf{D}\mathbf{a}_i, \sigma^2\right) + \log p(\mathbf{a}_i) \right] p\left(\mathbf{a}_i | \mathbf{x}_i, \Theta^{(l)}\right) d\mathbf{a}_i.$$

The dictionary is updated in the M-step as

$$\mathbf{D}^{(l+1)} = \underset{\mathbf{D}}{\operatorname{argmax}} \; \mathcal{Q}(\Theta|\Theta^{(l)}),$$

$$= \underset{\mathbf{D}}{\operatorname{argmax}} \sum_{i=1}^{T} \int_{\mathbf{a}_i} \left[ \log \mathcal{N} \left( \mathbf{x}_i | \mathbf{D}\mathbf{a}_i, \sigma^2 \right) + \log p(\mathbf{a}_i) \right] p \left( \mathbf{a}_i | \mathbf{x}_i, \Theta^{(l)} \right) d\mathbf{a}_i. \tag{3.11}$$

Assuming that each element in $\mathbf{a}_i$ follows an i.i.d. zero-mean Laplacian distribution, (3.11) can be expressed as

$$\mathbf{D}^{(l+1)} = \underset{\mathbf{D}}{\operatorname{argmin}} \sum_{i=1}^{T} \int_{\mathbf{a}_i} \left[ \|\mathbf{x}_i - \mathbf{D}\mathbf{a}_i\|_2^2 + \lambda \|\mathbf{a}_i\|_1 \right] p \left( \mathbf{a}_i | \mathbf{x}_i, \Theta^{(l)} \right) d\mathbf{a}_i, \tag{3.12}$$

where $\lambda$ combines all the constants in the exponents of the Gaussian and the Laplacian density functions. This is the general solution to a dictionary learning problem.

The deterministic formulation for computing the sparse codes and updating the dictionary in (3.2) can be obtained by constraining the posterior density $p(\mathbf{a}_i|\mathbf{x}_i, \Theta^{(l)})$, such that

$$p \left( \mathbf{a}_i | \mathbf{x}_i, \Theta^{(l)} \right) = \begin{cases} 1, & \text{if } \mathbf{a}_i = \hat{\mathbf{a}}_i, \\ 0, & \text{if } \mathbf{a}_i \neq \hat{\mathbf{a}}_i, \end{cases} \tag{3.13}$$

where $\hat{\mathbf{a}}_i$ is the minimum cost coefficient estimate computed by solving the $\ell_1$ minimization (2.4). The dictionary update now simplifies to (3.3). Note that the procedure obtained using the EM algorithm is the same as those obtained using alternating minimization of (3.2).

## 3.2   LEARNING ALGORITHMS

### 3.2.1   METHOD OF OPTIMAL DIRECTIONS

This algorithm was proposed by Engan *et al.* [96, 97], and is one of the early efforts for building learned dictionaries. The procedure optimizes (3.2) using alternating minimization, and, in the first step, the current estimate of the dictionary is used to solve for the sparse codes using the $\ell_0$ optimization

$$\min_{\mathbf{a}_i} \|\mathbf{x}_i - \mathbf{D}\mathbf{a}_i\|_2^2 \quad \text{subject to} \quad \forall i, \quad \|\mathbf{a}_i\|_0 \leq S, i = \{1, 2, \ldots, T\}, \tag{3.14}$$

which can be solved using algorithms such as OMP or FOCUSS [100], [101].

In the second step, the dictionary in the optimization is updated, assuming that the sparse codes are known. For the dictionary update step, we define the errors $\mathbf{e}_i = \mathbf{x}_i - \mathbf{D}\mathbf{a}_i$. Hence the MSE of the overall representation is given by

$$\text{MSE} = \frac{1}{T} \|\mathbf{X} - \mathbf{D}\mathbf{A}\|_F^2. \tag{3.15}$$

The algorithm attempts to update $\mathbf{D}$ such that the above error is minimized. Differentiating (3.15) with respect to $\mathbf{D}$, we obtain $(\mathbf{X} - \mathbf{D}\mathbf{A})\mathbf{A}^T = 0$. The dictionary update step can be expressed as

$$\mathbf{D}^{(l+1)} = \mathbf{X}\mathbf{A}^{(l)^T}(\mathbf{A}^{(l)}\mathbf{A}^{(l)^T})^{-1}, \tag{3.16}$$

where the superscript indicates the iteration number. The columns of the updated dictionary $\mathbf{D}^{(l+1)}$ are then normalized. Note that this is the best possible dictionary that can be obtained for the known coefficient matrix. Using other update schemes, such as iterative steepest descent, will result in a much slower learning procedure.

### 3.2.2  K-SVD

The next algorithm that we will consider is the K-SVD, proposed by Aharon *et al.* [95], which is also an alternating optimization procedure that attempts to minimize the objective in (3.2). Similar to the MOD, the dictionary $\mathbf{D}$ is fixed and the best coefficient matrix $\mathbf{A}$ is computed using a pursuit method, during the sparse coding step. The K-SVD algorithm has a markedly different dictionary update step, where it updates one column at a time, fixing all columns in the dictionary except one, $\mathbf{d}_k$, and updating it along with its corresponding coefficients such that the MSE is reduced. Modifying the coefficients during the dictionary update step imparts a significant acceleration in learning.

During the dictionary update, we assume that both $\mathbf{D}$ and $\mathbf{A}$ are fixed, except for one column in the dictionary, $\mathbf{d}_k$ and the coefficients corresponding to it, i.e., the $i^{\text{th}}$ row in $\mathbf{A}$ denoted by $\mathbf{a}_T^i$. Now, the primary objective function can be simplified as,

$$
\begin{aligned}
\|\mathbf{X} - \mathbf{D}\mathbf{A}\|_F^2 &= \left\| \mathbf{X} - \sum_{j=1}^{K} \mathbf{d}_j \mathbf{a}_T^j \right\|_F^2 \\
&= \left\| \mathbf{X} - \left( \sum_{j \neq k} \mathbf{d}_j \mathbf{a}_T^j \right) - \mathbf{d}_k \mathbf{a}_T^k \right\|_F^2 \\
&= \left\| \mathbf{E}_k - \mathbf{d}_k \mathbf{a}_T^k \right\|_F^2 .
\end{aligned}
\tag{3.17}
$$

It can be seen from (3.17) that the product term $\mathbf{D}\mathbf{A}$ has been decomposed into $K$ rank-1 matrices among which $K - 1$ terms are fixed. The matrix $\mathbf{E}_k$ indicates the errors for all the $T$ examples when the contribution of the $k^{\text{th}}$ atom is removed. The natural idea will be to use the SVD to find the new $\mathbf{d}_k$ and the corresponding $\mathbf{a}_T^k$. The SVD finds the closest rank-1 matrix, in terms of Frobenius norm, that will approximate $\mathbf{E}_k$ and thereby minimizes the expression in (3.17). However, this ignores the sparsity in $\mathbf{a}_T^k$ and will result in a dense representation.

The K-SVD algorithm handles this issue by defining a set $\omega_i$ containing the group of indices, pointing to the columns $\mathbf{x}_i$ of the input signal matrix, that use the atom $\mathbf{d}_k$. In effect,

$$\omega_k = \{i \mid 1 \leq i \leq K, \mathbf{a}_T^k[i] \neq 0\}. \tag{3.18}$$

**Table 3.1:** K-SVD Algorithm

**Initialization:**
Initial dictionary, $\mathbf{D}^{(0)}$, with normalized columns, $l = 1$.
**while** convergence not reached
  *Sparse Coding Step:*
    $\min_{\mathbf{a}_i} \{\|\mathbf{x}_i - \mathbf{D}\mathbf{a}_i\|_2^2\}$  subject to  $\|\mathbf{a}_i\|_0 \leq S$, for $i = \{1, \ldots, N\}$
  *Dictionary Update Step:* For each column $k$ in $\mathbf{D}^{(l-1)}$, update it by
  —Define $\omega_k = \{i \,|\, 1 \leq i \leq K, \mathbf{a}_T^k \neq 0\}$.
  —Compute $\mathbf{E}_k = \mathbf{X} - \left( \sum_{j \neq k} \mathbf{d}_j \mathbf{a}_T^j \right)$.
  —Compute $\mathbf{E}_R^k$, by selecting only the columns corresponding to $\omega_k$ from $\mathbf{E}_k$.
  —Apply SVD decomposition $\mathbf{E}_R^k = \mathbf{U} \boldsymbol{\Delta} \mathbf{V}^T$.
  —Choose the updated dictionary atom, $\mathbf{d}_k$, as the first column of $\mathbf{U}$.
  —Updated coefficient vector as the product of first column of $\mathbf{V}$ and $\boldsymbol{\Delta}(1,1)$.
  —Update the loop counter, $l = l + 1$.
**end**

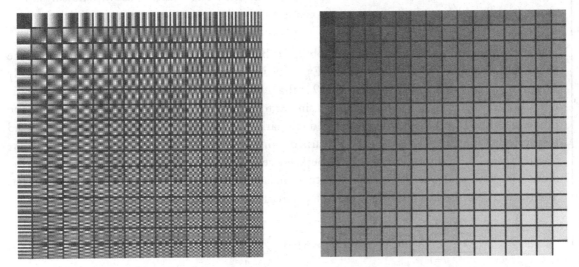

**Figure 3.2:** An overcomplete DCT dictionary (left) is shown along with the activity measures of its atoms (right), which is low for geometric atoms and high for texture-like atoms.

Now, define a matrix $\Omega_k$ with ones on the $(\omega_k[i], i)^{\text{th}}$ entries and zeros elsewhere. The vector $\mathbf{a}_R^k = \mathbf{a}_T^k \Omega_k$ contains only the non-zero entries in $\mathbf{a}_T^k$. Similarly, multiplying $\mathbf{X}_R^k = \mathbf{X}\Omega_k$ creates a matrix with the subset of the training examples that use the atom $\mathbf{d}_k$. The same happens in $\mathbf{E}_R^k = \mathbf{E}_k \Omega_k$. This penalty in (3.17) is simplified as

$$\left\| \mathbf{E}_k \Omega_k - \mathbf{d}_k \mathbf{a}_T^k \Omega_k \right\|_F^2 = \left\| \mathbf{E}_R^k - \mathbf{d}_k \mathbf{a}_R^k \right\|_F^2, \tag{3.19}$$

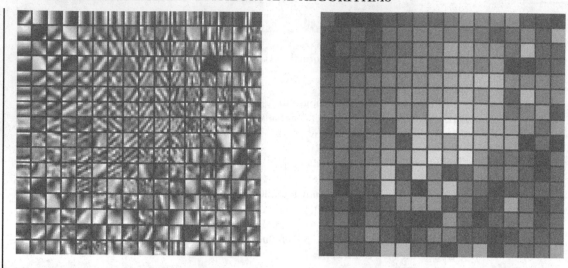

**Figure 3.3:** An example K-SVD dictionary (left) is shown along with the activity measures of its atoms (right), which is low for geometric atoms and high for texture-like atoms.

and the minimum of this expression is obtained by updating the dictionary atom $\mathbf{d}_k$ and the coefficient vector $\mathbf{a}_R^k$ using the SVD of $\mathbf{E}_R^k$. The steps involved in this algorithm are formally summarized in Table 3.1. It is shown in [95] that convergence of this algorithm to a local minimum is always guaranteed with a behavior similar to Gauss-Seidel methods in optimization. The predefined DCT and the learned K-SVD dictionaries are compared in Figure 3.2 and Figure 3.3 respectively. The K-SVD dictionary is learned from 50,000 grayscale patches of size $8 \times 8$ extracted from 250 training images of the Berkeley segmentation dataset (BSDS) [1]. Besides the dictionary atoms, a total-variation-like activity measure [20] also is shown for both dictionaries. The measure for the atom $\mathbf{d}$ organized as a patch of size $\sqrt{M} \times \sqrt{M}$ is defined as

$$\sum_{m=2}^{\sqrt{M}} \sum_{n=1}^{\sqrt{M}} |d[m,n] - d[m-1,n]| + \sum_{m=1}^{\sqrt{M}} \sum_{n=2}^{\sqrt{M}} |d[m,n] - d[m,n-1]|. \tag{3.20}$$

The measure is high for texture-like patches and low if the patches have a regular geometric structure. The measure is normalized such that its maximum value is 1 and shown for both DCT and K-SVD atoms in Figures 3.2 and 3.3. It can be seen that the DCT dictionary is more regular with respect to this measure compared to the K-SVD dictionary.

### 3.2.3    MULTILEVEL DICTIONARIES

The two learning algorithms described so far are generic, in the sense that they can be used for learning dictionaries from any training dataset, with the only *a priori* knowledge that they are sparsely representable in some overcomplete set of atoms. However, since the major application

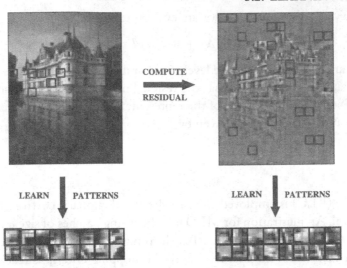

**Figure 3.4:** Features learned at two levels from non-overlapping patches (8 × 8) of a 128 × 96 image. The patches that are highlighted in the image and its residual share similar information and, hence, jointly correspond to a learned pattern (also highlighted).

of dictionary learning has been in image processing applications, we will describe the multilevel learning procedure that is especially suited for learning *global dictionaries* that are suitable for natural image representation. Global dictionaries are those that can provide a good representation for a wide range of test images, and such dictionaries learned from a set of randomly chosen patches from natural images have been successfully used for denoising [24], compressed sensing [102], and classification [38]. However, multilevel learning makes learning "good" global dictionaries an important goal in itself. This section will describe the learning algorithm and the theoretical properties of multilevel dictionaries will be discussed later in the chapter.

The motivation for multilevel learning is that natural images exhibit redundancy across local regions and there is also a clear hierarchy of patterns with respect to energy contribution to the patches. In addition to exhibiting redundancy, the natural image patches typically contain either geometric patterns or stochastic textures or a combination of both [103]. When dictionaries are learned in multiple levels according to the order of their energy contribution, geometric patterns that contribute higher energy are learned in the first few levels and stochastic textures are learned in the last few levels. From the algorithmic perspective, we move away from the paradigm of alternating optimization of a *flat dictionary*, to learning a *hierarchical dictionary* in just one iteration.

We denote the Multilevel Dictionary (MLD) as $\mathbf{D} = [\mathbf{D}_1 \ \mathbf{D}_2 \ \ldots \ \mathbf{D}_L]$, and the coefficient matrix as $\mathbf{A} = [\mathbf{A}_1^T \ \mathbf{A}_2^T \ \ldots \ \mathbf{A}_L^T]^T$. Here, $\mathbf{D}_l$ is the sub-dictionary in level $l$ and $\mathbf{A}_l$ is the coefficient matrix in level $l$. The approximation in level $l$ is expressed as

$$\mathbf{R}_{l-1} = \mathbf{D}_l \mathbf{A}_l + \mathbf{R}_l, \text{ for } l = \{1, \ldots, L\}, \tag{3.21}$$

where $\mathbf{R}_{l-1}$, $\mathbf{R}_l$ are the residuals for the levels $l-1$ and $l$ respectively and $\mathbf{R}_0 = \mathbf{X}$, the matrix of training image patches. This implies that the residual matrix in level $l-1$ serves as the training data for level $l$. Note that the sparsity of the representation in each level is fixed at 1. Hence, the overall approximation for all levels is given by

$$\mathbf{X} = \sum_{l=1}^{L} \mathbf{D}_l \mathbf{A}_l + \mathbf{R}_L. \tag{3.22}$$

K-lines clustering (KLC) is employed to learn the sub-dictionary $\mathbf{D}_l$ from the training matrix, $\mathbf{R}_{l-1}$, for that level. An illustration for MLD learning using patches of size $8 \times 8$ extracted from a $128 \times 96$ image is provided in Figure 3.4. MLD learning can be formally stated as an optimization problem that proceeds from the first level until the stopping criteria is reached. For level $l$, the optimization problem is

$$\underset{\mathbf{D}_l, \mathbf{A}_l}{\arg\min} \|\mathbf{R}_{l-1} - \mathbf{D}_l \mathbf{A}_l\|_F^2 \text{ subject to } \|\mathbf{a}_{l,i}\|_0 \leq 1,$$
$$\text{for } i = \{1, \ldots, T\}, \tag{3.23}$$

where $\mathbf{a}_{l,i}$ is the $i^{\text{th}}$ column of $\mathbf{A}_l$ and $T$ is the number of columns in $\mathbf{A}_l$. We adopt the notation $\{\mathbf{D}_l, \mathbf{A}_l\} = \text{KLC}(\mathbf{R}_{l-1}, K_l)$ to denote the problem in (3.23), where $K_l$ is the number of atoms in the sub-dictionary $\mathbf{D}_l$. The stopping criteria is provided by imposing a limit either on the residual representation error or the maximum number of levels ($L$). Note that the total number of levels is the same as the maximum number of non-zero coefficients (sparsity) of the representation. The error constraint can be stated as, $\|\mathbf{r}_{l,i}\|_2^2 \leq \epsilon, \forall i = 1, \ldots, T$ for some level $l$, where $\epsilon$ is the error goal.

Table 3.2 lists the MLD learning algorithm with sparsity and error constraints. We use the notation $\Lambda_l(j)$ to denote the $j^{\text{th}}$ element of the set $\Lambda_l$ and $\mathbf{r}_{l,i}$ denotes the $i^{\text{th}}$ column vector in the matrix $\mathbf{R}_l$. The set $\Lambda_l$ contains the indices of the residual vectors of level $l$ whose norm is greater than the error goal. The residual vectors indexed by $\Lambda_l$ are stacked in the matrix, $\hat{\mathbf{R}}_l$, which in turn serves as the training matrix for the next level, $l+1$. In MLD learning, for a given level $l$, the residual $\mathbf{r}_{l,i}$ is orthogonal to the representation $\mathbf{D}_l \mathbf{a}_{l,i}$. This implies that

$$\|\mathbf{r}_{l-1,i}\|_2^2 = \|\mathbf{D}_l \mathbf{a}_{l,i}\|_2^2 + \|\mathbf{r}_{l,i}\|_2^2. \tag{3.24}$$

Combining this with the fact that $\mathbf{y}_i = \sum_{l=1}^{L} \mathbf{D}_l \mathbf{a}_{l,i} + \mathbf{r}_{L,i}$, $\mathbf{a}_{l,i}$ is $1$−sparse, and the columns of $\mathbf{D}_l$ are of unit $\ell_2$ norm, we obtain the relation

$$\|\mathbf{x}_i\|_2^2 = \sum_{l=1}^{L} \|\mathbf{a}_{l,i}\|_2^2 + \|\mathbf{r}_{L,i}\|_2^2. \tag{3.25}$$

**Table 3.2:** Algorithm for building the multilevel dictionary

---

**Input**
$\mathbf{X} = [\mathbf{x}_i]_{i=1}^T$, $M \times T$ matrix of training vectors.
$L$, maximum number of levels of the dictionary.
$K_l$, number of dictionary elements in level $l$, $l = \{1, 2, ..., L\}$.
$\epsilon$, error goal of the representation.

**Output**
$\mathbf{D}_l$, adapted sub-dictionary for level $l$.

**Algorithm**
Initialize: $l = 1$ and $\mathbf{R}_0 = \mathbf{X}$.
$\Lambda_0 = \{i \mid \|\mathbf{r}_{0,i}\|_2^2 > \epsilon, 1 \leq i \leq T\}$, index of training vectors with
squared norm greater than error goal.
$\hat{\mathbf{R}}_0 = [\mathbf{r}_{0,i}]_{i \in \Lambda_0}$.

**while** $\Lambda_{l-1} \neq \emptyset$ and $l \leq L$
    Initialize:
        $\mathbf{A}_l$, coefficient matrix, size $K_l \times M$, all zeros.
        $\mathbf{R}_l$, residual matrix for level $l$, size $M \times T$, all zeros.
    $\{\mathbf{D}_l, \hat{\mathbf{A}}_l\} = \text{KLC}(\hat{\mathbf{R}}_{l-1}, K_l)$.
    $\mathbf{R}_l^t = \hat{\mathbf{R}}_{l-1} - \mathbf{D}_l\hat{\mathbf{A}}_l$.
    $\mathbf{r}_{l,i} = \mathbf{r}_{l,j}^t$ where $i = \Lambda_{l-1}(j)$, $\forall j = 1, ..., |\Lambda_{l-1}|$.
    $\mathbf{a}_{l,i} = \hat{\mathbf{a}}_{l,j}$ where $i = \Lambda_{l-1}(j)$, $\forall j = 1, ..., |\Lambda_{l-1}|$.
    $\Lambda_l = \{i \mid \|\mathbf{r}_{l,i}\|_2^2 > \epsilon, 1 \leq i \leq T\}$.
    $\hat{\mathbf{R}}_l = [\mathbf{r}_{l,i}]_{i \in \Lambda_l}$.
    $l \leftarrow l + 1$.
**end**

---

Equation (3.25) states that the energy of any training vector is equal to the sum of squares of its coefficients and the energy of its residual. From (3.24), we also have that

$$\|\mathbf{R}_{l-1}\|_F^2 = \|\mathbf{D}_l\mathbf{A}_l\|_F^2 + \|\mathbf{R}_l\|_F^2. \tag{3.26}$$

A multilevel dictionary, along with the activity measure of its atoms, given by (3.20), learned using the algorithm with the same training set as the K-SVD dictionary obtained in Section 3.2.2, is shown in Figure 3.5. The levelwise representation energy and the average activity measure for each level for the learned MLD are given in Figure 3.6, which clearly shows the energy hierarchy of the learning algorithm. This also demonstrates that the algorithm learns geometric patterns in

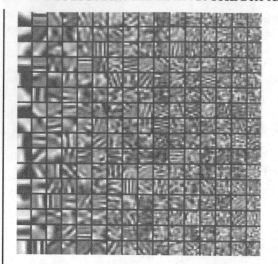 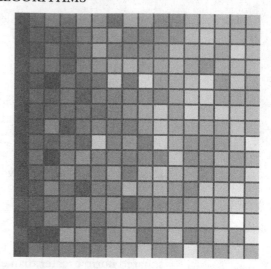

**Figure 3.5:** An example MLD with 16 levels of 16 atoms each (left), with the leftmost column indicating atoms in level 1, proceeding toward level 16 in the rightmost column. The dictionary comprises of geometric patterns in the first few levels, stochastic textures in the last few levels, and a combination of both in the middle levels, as quantified by its activity measure (right).

the first few levels, stochastic textures in the last few levels, and a hybrid set of patterns in the middle levels.

### Sparse Approximation using an MLD

Sparse approximation for any test data can be performed by stacking all the levels of an MLD together into a single dictionary, and using any standard pursuit algorithm on $\mathbf{D}$. Though this implementation is straightforward, it does not exploit the energy hierarchy observed in the learning process. The Regularized Multilevel OMP (RM-OMP) procedure incorporates energy hierarchy in the pursuit scheme by evaluating a 1-sparse representation for each level $l$ using the sub-dictionary $\tilde{\mathbf{D}}_l$, and orthogonalizes the residual to the dictionary atoms chosen so far. In order to introduce flexibility in the representation, this sub-dictionary is built using atoms selected from the current level as well as the $u$ immediately preceding and following levels, i.e., $\tilde{\mathbf{D}}_l = \left[\mathbf{D}_{l-u}\mathbf{D}_{l-(u-1)}\ldots\mathbf{D}_l\ldots\mathbf{D}_{l+(u-1)}\mathbf{D}_{l+u}\right]$. Considering an MLD with $L$ levels and $K/L$ atoms per level, the complexity of choosing $S$ dictionary atoms using the M-OMP algorithm is $S(2u + 1)KM/L$, where $M$ is the dimensionality of the dictionary atom. In contrast, the complexity of dictionary atom selection for the OMP algorithm is $SKM$, for a dictionary of size $M \times K$. If $u$ is chosen to be much smaller than $L$, we have $(2u + 1)/L < 1$, and, hence, the savings in computations for the RM-OMP algorithm is still significant in comparison to the standard OMP algorithm.

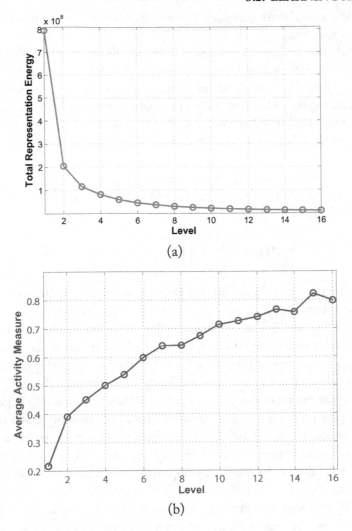

**Figure 3.6:** (a) Level-wise representation energy for the learned MLD with the BSDS training data set. (b) The levelwise activity measure shows that the atoms slowly change from geometric to texture-like as the level progresses.

## 3.2.4 ONLINE DICTIONARY LEARNING

The dictionary learning algorithms described so far are batch procedures, since they require access to the whole training set in order to minimize the cost function. Hence, the size of the data set that can be used for training is limited. In order for the learning procedure to scale up to millions of training samples, we need an online algorithm that is efficient, both in terms of memory and computations. We will discuss one such procedure proposed by Mairal *et al.* [104], based on stochastic approximations.

Although dictionary learning results in a solution that minimizes the *empirical cost g* given in (3.2), it will approach the solution that minimizes the *expected cost ĝ* given in (3.1), as the number of samples $T \to \infty$. The idea of online learning is to use a well-designed stochastic gradient algorithm that can result in a lower expected cost, when compared to an accurate empirical minimization procedure [105]. Furthermore, for large values of $T$, empirical minimization of (3.2), using a batch algorithm, becomes computationally infeasible and it is necessary to use an online algorithm.

The online algorithm alternates between computing sparse code for the $t^{th}$ training sample, $\mathbf{x}_t$, with the current estimate of the dictionary $\mathbf{D}_{t-1}$, and updating the new dictionary $\mathbf{D}_t$ by minimizing the objective

$$\hat{g}_t(\mathbf{D}) \equiv \frac{1}{t} \sum_{i=1}^{t} \left( \frac{1}{2} \|\mathbf{x}_i - \mathbf{D}\mathbf{a}_i\|_2^2 + \lambda \|\mathbf{a}_i\|_1 \right), \tag{3.27}$$

along with the constraint that the columns of $\mathbf{D}$ are of unit $\ell_2$ norm. The sparse codes for the samples $i < t$ are used from the previous steps of the algorithm. The online algorithm for dictionary learning is given in Table 3.3. The dictionary update using warm restart is performed by updating each column of it separately. The detailed algorithm for this update procedure is available in [104]. The algorithm is quite general and can be tuned to certain special cases. For example, when a fixed-size dataset is used, we can randomly cycle through the data multiple times and also remove the contribution of the previous cycles. When the dataset is huge, the computations can be speeded up by processing in mini-batches instead of one sample at a time. In addition, to improve the robustness of the algorithm, unused dictionary atoms must be removed and replaced with a randomly chosen sample from the training set, in a manner similar to clustering procedures. Since the reliability of the dictionary improves over iterations, the initial iterations may be slowed by adjusting the step size and down-weighting the contributions of the previous data.

In [104], it is proved that this algorithm converges to a stationary point of the objective function. This is shown by proving that, under the assumptions of compact support for the data, convexity of the functions $\hat{g}_t$ and uniqueness of the sparse coding solution, $\hat{g}_t$ acts as a converging surrogate of $g$, as the total number of training samples $T \to \infty$.

## 3.2.5  LEARNING STRUCTURED SPARSE MODELS

The generative model in (1.1) assumes that the data is generated as a sparse linear combination of dictionary atoms. When solving an inverse problem, the observations are usually a degraded version of the data denoted as

$$\mathbf{z} = \boldsymbol{\Phi}\mathbf{x} + \mathbf{n}, \tag{3.28}$$

where $\boldsymbol{\Phi} \in \mathbb{R}^{N \times M}$ is the degradation operator with $N \leq M$, and $\mathbf{n} \sim \mathcal{N}(\mathbf{0}, \sigma^2 \mathbf{I}_N)$. Considering the case of images, $\mathbf{x}$ usually denotes an image patch having the sparse representation $\mathbf{D}\mathbf{a}$, and

**Table 3.3:** Online Dictionary Learning

---

**Initialization:**
 —$T$ training samples drawn from the random distribution $p(\mathbf{x})$.
 —Initial dictionary, $\mathbf{D}^{(0)} \in \mathbb{R}^{M \times K}$, with normalized columns.
 —$\lambda$, regularization parameter.
 —Set $\mathbf{B} \in \mathbb{R}^{K \times K}$ and $\mathbf{C} \in \mathbb{R}^{M \times K}$ to zero matrices.
**for** $t = 1$ **to** $T$
 —Draw the training sample $\mathbf{x}_t$ from $p(\mathbf{x})$.
 —Sparse Coding:
$$\mathbf{a}_t = \operatorname{argmin}_{\mathbf{a}} \tfrac{1}{2}\|\mathbf{x}_i - \mathbf{D}_{t-1}\mathbf{a}\|_2^2 + \lambda\|\mathbf{a}\|_1$$
 —$\mathbf{B}_t = \mathbf{B}_{t-1} + \mathbf{a}_t \mathbf{a}_t^T$
 —$\mathbf{C}_t = \mathbf{C}_{t-1} + \mathbf{x}_t \mathbf{a}_t^T$
 —Compute $\mathbf{D}_t$ using $\mathbf{D}_{t-1}$ as warm restart,
$$\mathbf{D}_t = \tfrac{1}{t}\sum_{i=1}^{t}\left(\tfrac{1}{2}\|\mathbf{x}_i - \mathbf{D}\mathbf{a}_i\|_2^2 + \lambda\|\mathbf{a}_i\|_1\right)$$
$$= \tfrac{1}{t}\sum_{i=1}^{t}\left(\tfrac{1}{2}\operatorname{Tr}(\mathbf{D}^T\mathbf{D}\mathbf{A}_t) - \operatorname{Tr}(\mathbf{D}^T\mathbf{D}\mathbf{A}_t)\right)$$
**end for**
 —Return learned dictionary $\mathbf{D}_T$.

---

the matrix $\boldsymbol{\Phi}$ performs operations such as downsampling, masking, or convolution. Restoration of original images corrupted by these degradations corresponds to the inverse problems of single-image superresolution, inpainting, and deblurring respectively. The straightforward method of restoring these images, using sparse representations, is to consider $\boldsymbol{\Phi}\mathbf{D}$ as the dictionary and compute the sparse coefficient vector $\mathbf{a}$ using the observation $\mathbf{z}$. The restored data is then given as $\mathbf{D}\mathbf{a}$. Apart from the obvious necessity that $\mathbf{x}$ should be sparsely decomposable in $\mathbf{D}$, the conditions to be satisfied in order to get a good reconstruction are [106]: (a) the norms of any column in the dictionary $\boldsymbol{\Phi}\mathbf{D}$ should not become close to zero, as in this case it is not possible to recover the corresponding coefficient with any confidence; and (b) the columns of $\boldsymbol{\Phi}\mathbf{D}$ should be incoherent, since coherent atoms lead to ambiguity in coefficient support. For uniform degradations such as downsampling and convolution, even an orthonormal $\mathbf{D}$ could result in $\boldsymbol{\Phi}\mathbf{D}$ that violate these two conditions. For example, consider that $\boldsymbol{\Phi}$ is an operator that downsamples by a factor of two, in which case the DC atom $\{1, 1, 1, 1, \ldots\}$ and the highest frequency atom $\{1, -1, 1, -1, \ldots\}$ will become identical after downsampling. If the dictionary $\mathbf{D}$ contains directional filters, an isotropic degradation operator $\boldsymbol{\Phi}$ will not lead to a complete loss in the incoherence property for $\boldsymbol{\Phi}\mathbf{D}$.

In order to overcome these issues and obtain a stable representation for inverse problems, it is necessary to consider the fact that similar data admit similar representations and design dictionaries, accordingly. As discussed in Chapter 2, this corresponds to the simultaneous sparse approximation (SSA) problem where a set of data will be represented by a set of dictionary atoms. In the section, we will briefly discuss the formulations by Yu *et al.* [106] and Mairal *et al.* [107] that are specifically designed for inverse problems in imaging.

**Dictionaries for Simultaneous Sparse Approximation**

We restate the problem of SSA discussed in Chapter 2 and elaborate on it from a dictionary learning perspective. Let us assume that the data samples given by $\{\mathbf{x}_i\}_{i=1}^{T}$ can be grouped into $G$ disjoint sets, and the indices of the members of group $g$ are available in $\mathcal{A}_g$. Note that the each dictionary atom can participate in more than one group of representations. The problem of SSA and dictionary learning is given by

$$\min_{\mathbf{D},\mathbf{A}} \|\mathbf{X} - \mathbf{D}\mathbf{A}\|_F^2 + \lambda \sum_{g=1}^{G} \|\mathbf{A}_g\|_{p,q}. \tag{3.29}$$

The matrix $\mathbf{A}_g \in \mathbb{R}^{K \times |\mathcal{A}_g|}$ contains the coefficient vectors of data samples that belong to the group $g$. The pseudonorm $\|\mathbf{A}\|_{p,q}$ is defined as $\sum_{i=1}^{K} \|\mathbf{a}^i\|_q^p$, where $\mathbf{a}^i$ denotes the $i^{\text{th}}$ row of $\mathbf{A}$ and the pair $(p, q)$ is chosen to be $(1, 2)$ or $(0, \infty)$. The choice of $p = 1$ and $q = 2$ results in a convex penalty. Dictionary learning can be carried out using an alternating minimization scheme, where the dictionary is obtained by fixing the coefficients and the coefficients are obtained by fixing the dictionary. In practice, two basic issues need to be addressed in the group sparse coding stage that is usually carried out using an error constrained optimization: (a) how should the groups be selected? and (b) what is the error goal for the representation?

We provide simple answers to these questions that lead to efficient implementation schemes, following [107]. Patches can be grouped by clustering them into a prespecified number of groups, thereby keeping the similar patches together and avoiding the problem of overlapping in groups. The error goal for the group, $\epsilon_g$, is the goal for one patch scaled by the number of elements in the group. Similar to BPDN in (2.21), the error constrained minimization is now obtained as

$$\min_{\{\mathcal{A}_g\}_{g=1}^{G}} \sum_{g=1}^{G} \frac{\|\mathbf{A}_g\|_{p,q}}{|\mathcal{A}_g|^p} \text{ such that } \sum_{i \in \mathcal{A}_g} \|\mathbf{x}_i - \mathbf{D}\mathbf{a}_i\|_2^2 \leq \epsilon_g, \tag{3.30}$$

where the group norm is normalized to ensure equal weighting for all the groups. When the sparse codes are fixed, dictionary learning can be performed using any standard procedure. With denoising and demosaicing applications, a proper adaptation of the SSA and dictionary learning procedure leads to state-of-the-art results.

**Dictionaries for Piecewise Linear Approximation**

The main idea in building these dictionaries is to design a set of directional bases, and represent each cluster of image patches linearly using a basis set. Linear modeling, using directional basis and simultaneous approximation, impart stability to the representation, and, hence, ensures that the restoration performance is high for the degraded data given by (3.28). The patches themselves are modeled as a mixture of $C$ Gaussians, each representing a linear model, i.e., $p(\mathbf{x}_i) = \mathcal{N}(\boldsymbol{\mu}_{c_i}, \boldsymbol{\Sigma}_{c_i})$. An EM-MAP procedure computes the *Maximum-a-Posteriori* (MAP) estimates for the restored patches, using an *Expectation-Maximization* (EM) algorithm, by per-

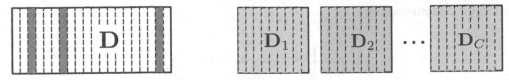

**Figure 3.7:** Sparse coding with dictionaries (left) and piecewise linear approximation using multiple PCA bases (right).

forming the following steps iteratively: (a) given a set of degraded patches $\{\mathbf{z}_i\}_{i=1}^{T}$, the parameters of the $C$ Gaussians are obtained using maximum likelihood estimation; (b) for each patch $i$, identify the Gaussian that generates it; and (c) estimate each non-degraded patch $\mathbf{x}_i$ from its corresponding Gaussian. Without loss of generality, we can assume that the means $\{\mu_c\}_{c=1}^{C}$ are zero, and, hence, the only parameter to be estimated for each Gaussian is its covariance $\Sigma_j$. In order to compute the non-degraded patch and its membership with respect to a cluster, we perform the following MAP estimation:

$$\{\mathbf{x}_i, c_i\} = \underset{\mathbf{x},c}{\operatorname{argmax}} \log p(\mathbf{x}|\mathbf{z}_i, \Sigma_c).$$

Using Bayes rule and realizing that $\mathbf{n} \sim \mathcal{N}(\mathbf{0}, \sigma^2 \mathbf{I}_M)$, we have

$$\{\mathbf{x}_i, c_i\} = \underset{\mathbf{x},c}{\operatorname{argmax}} \left[ \log p(\mathbf{z}_i|\mathbf{x}, \Sigma_c) + \log p(\mathbf{x}|\Sigma_c) \right], \tag{3.31}$$

$$= \underset{\mathbf{x},c}{\operatorname{argmin}} \left[ \|\mathbf{U}_i\mathbf{x} - \mathbf{z}_i\|_2^2 + \sigma^2 \mathbf{x}^T \Sigma_c^{-1}\mathbf{x} + \sigma^2 \log|\Sigma_c| \right]. \tag{3.32}$$

This joint estimation can be divided into a computation of per-cluster data $\mathbf{x}_i^c$ and the cluster membership $c_i$ as follows,

$$\mathbf{x}_i^c = \underset{\mathbf{x}}{\operatorname{argmin}} \left[ \|\mathbf{U}_i\mathbf{x} - \mathbf{z}_i\|_2^2 + \sigma^2 \mathbf{x}^T \Sigma_c^{-1}\mathbf{x} \right], \tag{3.33}$$

$$c_i = \underset{c}{\operatorname{argmin}} \left[ \|\mathbf{U}_i\mathbf{x}_i^c - \mathbf{z}_i\|_2^2 + \sigma^2 (\mathbf{x}_i^c)^T \Sigma_c^{-1}\mathbf{x}_i^c + \sigma^2 \log|\Sigma_c| \right], \tag{3.34}$$

and finally assigning $\mathbf{x}_i = \mathbf{x}_i^c$. Since $\Sigma_c$ could become rank-deficient, a small constant is added to it in order to regularize it. This procedure is also referred to as Piecewise Linear Estimation (PLE) since each Gaussian in the mixture corresponds to a linear model.

In order to provide a structured sparse model interpretation and to overcome the issue of rank deficiency in $\Sigma_c$, we perform the eigen decomposition, $\Sigma_c = \mathbf{D}_c \Lambda_c \mathbf{D}_c^T$, where $\Lambda_c$ is a diagonal matrix with the decreasing eigenvalues $\{\lambda_1^c, \ldots \lambda_M^c\}$. The patch belonging to that Gaussian can be conveniently represented using the obtained PCA basis as

$$\mathbf{x}_i^c = \mathbf{D}_c \mathbf{a}_i^c. \tag{3.35}$$

Hence, (3.33) transforms to

$$\mathbf{a}_i^c = \underset{\mathbf{a}}{\text{argmin}} \left( \|\mathbf{\Phi}_i \mathbf{D}_c \mathbf{a} - \mathbf{z}_i\|_2^2 + \sigma^2 \sum_{m=1}^{M} \frac{|\mathbf{a}(m)|^2}{\lambda_m^c} \right), \tag{3.36}$$

which is a linear estimation problem on the basis $\mathbf{D}_c$. Note that this is similar to a SSA problem, assuming that the group of dictionary atoms that represent the data are known. The selection of an optimal basis for the data $\mathbf{x}_i$ is equivalent to the selection of the best cluster in (3.34). Therefore, the EM-MAP procedure can be implemented by iteratively performing: (a) linear estimation in the $C$ basis sets, each of which corresponds to a cluster; (b) selection of the best basis set for each patch; and (c) estimating the basis sets using the associated patches. The eigenvalues for any cluster fall rapidly, and, hence, (3.36) leads to a very sparse representation. In actual implementation, (3.36) is never used to estimate the non-degraded patch. Instead, (3.33) and (3.34) are used to perform recovery of the degraded patches.

Since the PLE procedure is non-convex, the algorithm can only converge to a local minimum. Therefore, initialization of the model plays a key role in ensuring a good estimate of $\mathbf{x}$. The initialization also depends on the degradation operator $\mathbf{\Phi}$, since the recoverability condition that the norms of columns of $\mathbf{\Phi}\mathbf{D}$ be greater than zero has to be satisfied. For operators such as downsampling and random masking, the initialization is performed with a *directional PCA* basis as follows: images with oriented edges at different angles are chosen and patches are extracted from different positions in the edge regions; and PCA is performed on the extracted patches to compute the basis and the eigen values. They are shown in Figure 3.8, and correspond to Diracs in Fourier space. In practice, the eigen values are computed only once and the eigen vectors are computed using a Gram Schmidt procedure to initialize the covariance matrices. When $\mathbf{\Phi}$ is a blur operator, directional PCA basis cannot be used for initialization since they will not satisfy the recoverability condition. Therefore, we compute the position PCA basis, that are spread in Fourier space, by blurring the edge image of a specific orientation with various amounts of blur and extracting the patches from the same positions. After initialization, the multiple iterations of the MAP-EM procedure is used with patches extracted from degraded images, in order to compute the restored image.

## 3.2.6   SPARSE CODING USING EXAMPLES

Apart from sparse representation models that represent a union of subspaces, low dimensional manifold models also have been very successful in the representation and processing of natural signals and images. Typically, we do not have access to a complete description of the manifold, rather we are only presented with samples obtained from it. Dictionaries learned from a sufficient number of training examples from the data manifold capture the key features present in the data. The efficiency of sparse-representation-based recovery schemes can be improved if additional regularization is performed using the higher dimensional training examples from the data

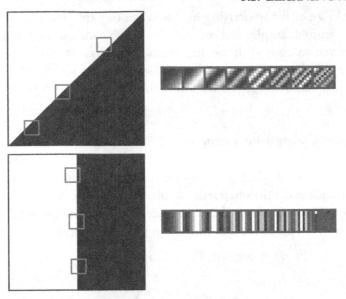

**Figure 3.8:** Examples for directional PCA basis.

manifold. This is particularly useful when the observations have highly reduced dimensions and are corrupted by a high degree of noise.

**Weighted Sparse Coding for Manifold Projection**

Let the set of vectors in $\mathbf{X}$ be the samples obtained from the manifold $\mathcal{M}$. Projecting a test sample $\mathbf{y}$ onto the manifold $\mathcal{M}$ can be performed by selecting a small set of samples $\mathbf{X}_{\Omega} = \{\mathbf{x}_i\}_{i \in \Omega}$ in the neighborhood of $\mathbf{y}$ and projecting $\mathbf{y}$ onto the simplex spanned by $\mathbf{X}_{\Omega}$. $\Omega$ is the index set of manifold samples in the neighborhood of $\mathbf{y}$. Usually the neighborhood is chosen using an $\ell_2$ distance measure—this approach has been used in manifold learning algorithms such as the Locally Linear Embedding (LLE) [108]. The selection of the neighborhood and projection of the test sample onto the low dimensional simplex can be posed as a weighted sparse coding problem,

$$\hat{\boldsymbol{\beta}} = \underset{\boldsymbol{\beta}}{\operatorname{argmin}} \|\mathbf{y} - \mathbf{X}\boldsymbol{\beta}\|_2^2 + \lambda \sum_{i=1}^{T} \|\mathbf{y} - \mathbf{x}_i\|_2^2 |\beta_i|,$$

$$\text{subject to } \sum_{i=1}^{T} \beta_i = 1, \beta_i \geq 0 \;\forall i, \tag{3.37}$$

where $\boldsymbol{\beta} \in \mathbb{R}^T$ are coefficients for the projection of $\mathbf{y}$ on $\mathbf{X}$. The constraints on $\boldsymbol{\beta}$ ensure that the test sample is projected onto the convex hull (simplex) spanned by the chosen manifold samples. The optimization problem given in (3.37) has been used in [39] to learn manifolds from high-dimensional data and shown to be highly beneficial for image classification.

The knowledge of the underlying manifold can be exploited by learning a dictionary directly from the manifold samples and computing a sparse code for the test data using the learned dictionary. However, in cases with missing/incomplete data, we have the observations $\mathbf{z} \in \mathbb{R}^N$ obtained as $\boldsymbol{\Phi}\mathbf{y} + \mathbf{n}$, where $\mathbf{n}$ denotes additive noise. When $N << M$ and $\sigma^2$ is high, apart from sparsity constraints, additional regularization in the form of samples from the manifold will aid in the recovery of $\mathbf{y}$.

**Regularizing Sparse Coding using Examples**

The goal here is to recover the unknown test sample $\mathbf{y}$ using both the sparsifying dictionary and examples from the manifold. The framework proposed in [50] involves two components: (a) computing the sparse code using the observation $\mathbf{z}$ and the dictionary $\mathbf{D}$; and (b) projection of the recovered test sample $\mathbf{Da}$ close to $\mathcal{M}$. This, in essence, involves combining the sparse representation and manifold projection problems into the joint optimization problem

$$\{\hat{\mathbf{a}}, \hat{\boldsymbol{\beta}}\} = \operatorname*{argmin}_{\mathbf{a}, \boldsymbol{\beta}} \|\mathbf{Da} - \mathbf{X}\boldsymbol{\beta}\|_2^2 + \lambda \|\mathbf{a}\|_1,$$

$$\text{subject to } \|\mathbf{z} - \boldsymbol{\Phi}\mathbf{Da}\|_2^2 \leq \epsilon, \sum_{i=1}^{T} \beta_i = 1, \beta_i \geq 0 \ \forall i. \tag{3.38}$$

Here, $\mathbf{a}$ denotes the sparse code for the observation $\mathbf{z}$, $\boldsymbol{\beta}$ corresponds to the coefficient vector obtained from manifold projection, and $\lambda$ controls the trade-off between sparsity of $\mathbf{a}$ and the manifold projection residual. The first term in the objective function, along with the second and third constraints, perform the manifold projection on the recovered data $\mathbf{Da}$. When compared to (3.37), the manifold projection component in the cost function of (3.38) does not contain the weighted sparsity term, $\sum_{i=1}^{T} \|\mathbf{z} - \boldsymbol{\Phi}\mathbf{x}_i\|_2^2 |\beta_i|$. This speeds up the optimization procedure and furthermore the constraints $\sum_{i=1}^{T} \beta_i = 1$ and $\beta_i \geq 0$ are strong enough to ensure a reasonable projection onto the data manifold. $\epsilon$ is the maximum residual error allowed for the sparse representation.

From the term $\|\mathbf{z} - \boldsymbol{\Phi}\mathbf{Da}\|_2^2 \leq \epsilon$, we can observe that the error ellipsoid for the sparse recovery is constrained only in $N$ dimensions as $\boldsymbol{\Phi}\mathbf{D} \in \mathbb{R}^{N \times K}$. When $N$ is not large enough to permit a good sparse approximation, constraining $\mathbf{Da}$ to be close to the manifold $\mathcal{M}$ will improve the approximation. A simplified version of this scenario is illustrated in Figure 3.9. We have $M = 2$ and $N = 1$ in this case. The test data $\mathbf{y}$ is lying close to the manifold $\mathcal{M}$. The noisy observation of $\mathbf{y}$ with one of the dimensions removed is $\mathbf{z}$. $\mathbf{e}_1$ and $\mathbf{e}_2$ are the canonical basis vectors. The first constraint in (3.38) specifies that the projection of the recovered data onto $\mathbf{e}_1$ should lie within $\sqrt{\epsilon}$ of $\mathbf{z}$. The shaded region indicates the possible locations of the recovered data. Incorporating the constraint that the recovered data $\mathbf{Da}$ is close to $\mathcal{M}$ improves the chance of $\mathbf{Da}$ being close to $\mathbf{y}$. This is particularly true when $N$ is quite small, in which case sparse coding alone cannot provide a good recovery for the test data from its noisy observation. We can implement (3.38) using standard convex optimization packages by defining a new variable $\boldsymbol{\eta} = [\mathbf{a}^T \ \boldsymbol{\beta}^T]^T$ and rewriting (3.38) in terms of this variable.

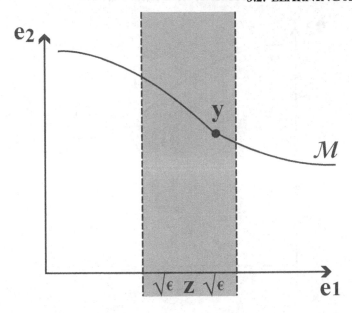

**Figure 3.9:** The data **y** lying near manifold $\mathcal{M}$ is corrupted with noise and one of the dimensions is removed to result in **z**. The shaded region indicates the possible locations of recovered data, subject to error constraints.

**Combining Sparse Coding and Manifold Projection**

Instead of regularizing the sparse code using manifold samples, we also can assume that the test data **y** has two components, one that can be represented by manifold projection onto the training examples **X**, and the other that can be well-represented using the pre-defined dictionary **D**. This model can be expressed as

$$\mathbf{y} = \mathbf{X}_{\Omega}\beta_{\Omega} + \mathbf{D}\mathbf{a}, \tag{3.39}$$

where $\Omega$ is the neighborhood of **y** assumed to be known. The only additional constraint posed here is $\beta_{\Omega} \geq 0$, since our experiments showed that the additional constraint that the elements of $\beta_{\Omega}$ sum to one did not yield a significant improvement in performance.

The test observation **z**, which is a noisy and low-dimensional version of **y** and this model, is illustrated in Figure 3.10 for a low-dimensional case when $M = 2$ and $N = 1$. Since the neighborhood $\Omega$ is usually not known, an estimated neighborhood $\hat{\Omega}$ is obtained by computing the distance between **z** and the low-dimensional examples $\boldsymbol{\Phi}\mathbf{X}$. The coefficients can be computed by solving the optimization program

$$\{\hat{\beta}_{\hat{\Omega}}, \hat{\mathbf{a}}\} = \operatorname*{argmin}_{\beta_{\hat{\Omega}}, \mathbf{a}} \|\beta_{\hat{\Omega}}\|_1 + \|\mathbf{a}\|_1 \text{ subject to } \|\mathbf{z} - \boldsymbol{\Phi}\mathbf{X}\beta_{\hat{\Omega}} - \boldsymbol{\Phi}\mathbf{D}\mathbf{a}\|_2^2 \leq \epsilon, \alpha_{\hat{\Omega}} \geq \mathbf{0}. \tag{3.40}$$

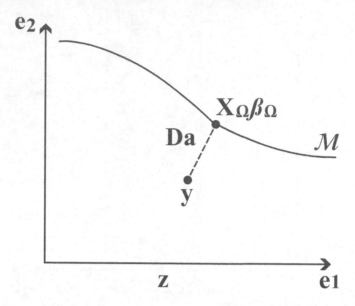

**Figure 3.10:** The test data **y** has a component that can be represented in the manifold $\mathcal{M}$ and another component that can be sparsely represented in a dictionary **D**. The test observation **z** is a noisy, low dimensional version of the test data **y**.

The above program can be solved using the COMB-BP algorithm or its greedy variant, the COMB-OMP, proposed in [109]. In both cases, we need to modify the algorithms to incorporate stopping criteria based on the error constraint. The COMB-BP can be efficiently implemented by modifying the LARS algorithm [74] to incorporate the non-negative constraint on a part of the coefficient vector.

## 3.3 STABILITY AND GENERALIZABILITY OF LEARNED DICTIONARIES

A learning algorithm is a map from the space of training examples to the hypothesis space of functional solutions. Algorithmic stability characterizes the behavior of a learning algorithm with respect to the perturbations of its training set [110], and generalization ensures that the expected error of the learned function, with respect to the novel test data, will be close to the average empirical training error [111]. Since clustering is a special case of dictionary learning, understanding the stability and generalization behavior of clustering algorithms is important for understanding these characteristics with respect to dictionary learning.

In clustering, the learned function is completely characterized by the cluster centers. Stability of a clustering algorithm implies that the cluster centroids learned by the algorithm are not significantly different when different sets of i.i.d. samples from the same probability space are used for training [112]. When there is a unique minimizer to the clustering objective with respect to the underlying data distribution, stability of a clustering algorithm is guaranteed [113], and this analysis has been extended to characterize the stability of K-means clustering in terms of the number of minimizers [114]. In [99], the stability properties of the K-lines clustering algorithm have been analyzed and have been shown to be similar to those of K-means clustering. Note that all the stability characterizations depend only on the underlying data distribution and the number of clusters, and not on the actual training data itself. Generalization implies that the average empirical training error becomes asymptotically close to the expected error, with respect to the probability space of data, as the number of training samples $T \to \infty$. In [115], the generalization bound for sparse coding in terms of the number of samples, $T$, also referred to as sample complexity, is derived and, in [116], improved by assuming a class of dictionaries that is nearly orthogonal.

## 3.3.1   EMPIRICAL RISK MINIMIZATION

Ideally, the parameters in a learning algorithm should be inferred based on the probability space of the training examples. This can be performed by posing a cost function and identifying the parameters that result in the minimum expected risk with respect to that cost function. However, since, in almost all the practical cases, the probability space of the training data is unknown, and we only have the samples realized from it, we resort to minimizing the average empirical cost instead of the expected cost. This procedure is referred to as Empirical Risk Minimization (ERM). Unsupervised clustering algorithms can also be posed as an ERM procedure by defining a hypothesis class of loss functions to evaluate the possible cluster configurations and to measure their quality [117]. For example, K-lines clustering can be posed as an ERM problem over the distortion function class

$$\mathcal{L}_K = \left\{ l_\mathbf{D}(\mathbf{x}) = \rho(\mathbf{x}, \mathbf{d}_j), j = \underset{l \in \{1,\dots,K\}}{\operatorname{argmax}} |\mathbf{x}^T d_l| \right\}, \tag{3.41}$$

where $\rho(\mathbf{x}, \mathbf{d}_j)$ is the distortion measure defined in (3.4). The class $\mathcal{L}_K$ is obtained by taking functions $l_D$ corresponding to all possible combinations of $K$ unit length vectors from the $\mathbb{R}^M$ space for the set $\mathbf{D}$. Let us formally define the probability space for the training data in $\mathbb{R}^M$ as $(\mathcal{X}, \Sigma, P)$, where $\mathcal{X}$ is the sample space and $\Sigma$ is a sigma-algebra on $\mathcal{X}$, i.e., the collection of subsets of $\mathcal{X}$ over which the probability measure $P$ is defined. The training samples, $\{\mathbf{x}_i\}_{i=1}^T$, are $T$ i.i.d. realizations from the probability space. Note that the stability of clustering algorithms that can be posed as ERM procedures depends heavily on the hypothesis class of loss functions. For instance, it has been shown that K-means and K-lines clustering are stable, if the probability space of the training data satisfy certain weak conditions.

## 3.3.2   AN EXAMPLE CASE: MULTILEVEL DICTIONARY LEARNING

In dictionary learning, algorithmic stability implies that, given a sufficiently large training set, the learning algorithm will result in global dictionaries that will depend only on the probability space to which the training samples belong and not on the actual samples themselves. Generalization ensures that such global dictionaries learned result in a good performance with test data. In other words, the asymptotic stability and generalization of a dictionary learning algorithm provides the theoretical justification for the uniformly good performance of global dictionaries learned from an arbitrary training set.

### Stability and Generalization Characteristics of MLD

The multilevel dictionary learning algorithm is shown to be asymptotically stable and generalizable [118]. Since MLD designs dictionaries using multiple levels of K-lines clustering, the stability characteristics of K-lines clustering can be used to prove the stability of MLD learning. If there is a unique minimizer to the clustering objective in all levels of MLD learning, then the MLD algorithm is stable even for completely disjoint training sets, as $T \to \infty$. However, if there are multiple minimizers in at least one level, the algorithm is stable only with respect to a change of $o(\sqrt{T})$ training samples between the two clusterings. In particular, a change in $\Omega(\sqrt{T})$ samples makes the algorithm unstable. In MLD learning, when the number of samples used for training $T \to \infty$, a generalizable dictionary is obtained. Therefore, when test data is coded using the RM-OMP coding scheme using this dictionary, the error in representation obtained will be similar to the empirical training error.

### Demonstration

Both the stability and generalization characteristics of MLD are crucial for building effective global dictionaries to model natural image patches. Although it is not possible to demonstrate the asymptotic behavior experimentally, the following simulations will illustrate the changes in the behavior of the learning algorithm with an increase in the number of samples used for training.

In order to illustrate the stability characteristics of MLD learning, we set up an experiment where we consider a multilevel dictionary of 4 levels, with 8 atoms in each level. We extracted grayscale patches of size $4 \times 4$ from the BSDS training images and trained multilevel dictionaries using various number of training patches. It has been shown for MLD [118] that asymptotic stability is guaranteed when the training set is changed by not more than $o(\sqrt{T})$ samples. In other words, the inferred dictionary atoms will not vary significantly, if this condition is satisfied.

We fixed the size of the training set at different values $T = \{1,000, 5,000, 10,000, 50,000, 100,000\}$ and learned an initial set of dictionaries using the proposed algorithm. The second set of dictionaries were obtained by replacing different number of samples from the original training set. For each case of $T$, the number of replaced samples was varied between 100 and $T$. For example, when $T = 10,000$, the number of replaced training samples were 100, 1,000, 5,000, and 10,000. The amount of change between the initial and the second set of dictionaries was quantified using

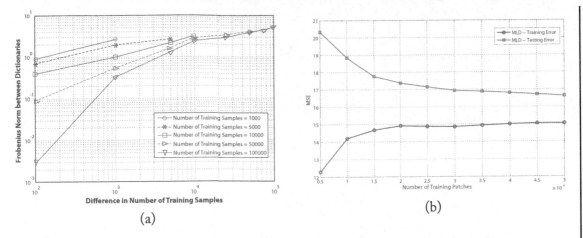

**Figure 3.11:** (a) Demonstration of the stability behavior of the proposed MLD learning algorithm. The minimum Frobenius norm between the difference of two dictionaries with respect to permutation of their columns and signs is shown. The second dictionary is obtained by replacing a different number of samples in the training set, used for training the original dictionary, with new data samples. (b) Demonstration of the generalization characteristics of the proposed algorithm. We plot the MSE obtained by representing patches from the BSDS test dataset, using dictionaries learned with a different number of training patches. For comparison, we show the training error obtained in each case.

the minimum Frobenius norm of their difference with respect to permutations of their columns and sign changes. In Figure 3.11(a), we plot this quantity for different values of $T$ as a function of the number of samples replaced in the training set. For each case of $T$, the difference between the dictionaries increases as we increase the replaced number of training samples. Furthermore, for a fixed number of replaced samples (say 100), the difference reduces with the increase in the number of training samples, since it becomes closer to asymptotic behavior.

Generalization of a dictionary learning algorithm guarantees a small approximation error for a test data sample, if the training samples are well approximated by the dictionary. In order to demonstrate the generalization characteristics of MLD learning, we designed dictionaries using different numbers of training image patches (grayscale), of size $8 \times 8$, obtained from the BSDS training dataset, and evaluated the sparse codes for patches obtained from 50 images in the BSDS test dataset. The dictionaries were of size $64 \times 256$ with 16 atoms per level. Figure 3.11(b) shows the approximation error (MSE) for both the training and test datasets, obtained using multilevel dictionaries. In all cases, the sparsity in training and testing was fixed at $S = 16$. As it can be observed, the difference between the MSE for training and test data reduces with the increase in the size of the training set, which validates our claim about generalization of MLD.

# CHAPTER 4

# Compressed Sensing

Most natural signals have low degrees of freedom compared to their ambient dimensions, yet the Shannon sampling theorem dictates that, in order to capture the signal without losing any information, a sampling rate of at least twice the bandwidth of the signal is required. Many times, a sensed signal is compressed and much of the redundancy is thrown away. The low-dimensionality of the signal can be exploited and significantly fewer numbers of samples can be acquired when preserving the information contained, using a proper sensing methodology.

Compressive sensing is a new sampling paradigm that attempts to infer the signal, or the necessary information from it, by asking a series of questions about the signal. In the well-known linear compressive sensing paradigm [119, 120], each question corresponds to a linear measurement of the signal using a measurement vector. Denoting the signal as $\mathbf{x} \in \mathbb{R}^M$ and the measurement matrix as $\boldsymbol{\Phi} \in \mathbb{R}^{N \times M}$, each row of the measurement matrix corresponds to a question, and the sensed signal is given by

$$\mathbf{z} = \boldsymbol{\Phi}\mathbf{x}, \tag{4.1}$$

where the number of measurements $N$ is less than the dimensions $M$. It is clear that, in order to perform recovery of this signal, some prior knowledge about the low-dimensionality of the signal is necessary, and, here, we will only consider the case where $\mathbf{x}$ is sparse in some basis, i.e., $\mathbf{x} = \mathbf{D}\mathbf{a}$ with $\mathbf{a}$ being the sparse vector. Note that $\mathbf{a}$ is assumed to be $K-$dimensional and therefore $\mathbf{D} \in \mathbb{R}^{M \times K}$. Although the sensing or encoding process is performed using the linear operator $\boldsymbol{\Phi}$, the decoding or reconstruction process is denoted by the nonlinear operator $\boldsymbol{\Delta}$ such that $\mathbf{x} = \boldsymbol{\Delta}(\boldsymbol{\Phi}\mathbf{x})$. Determining the appropriate encoder/decoder pair is of paramount importance in compressive sensing. In practice, we are interested in recovering signals that are not just sparse, but those that are *compressible* in a basis, i.e., those that have a few large coefficients and several small coefficients.

In order to understand why compressive sensing is important, let us consider coding an image using wavelet transform. Typically, it is possible to get a good reconstruction of an image using only 5% of the coefficients. Suppose we perform compressive sensing of the signal as described by $\mathbf{z} = \boldsymbol{\Phi}\mathbf{D}\mathbf{a}$. In order not to lose any information during the measurement process, the rows of the matrix $\boldsymbol{\Phi}$ must not be well-representable using the columns of $\mathbf{D}$. In other words, the measurement system must be completely unstructured as the wavelet basis represents the key structures present in natural images. This can be achieved by sampling the elements of $\boldsymbol{\Phi}$ from a random distribution. By using noise-like and incoherent measurements, it has been shown that

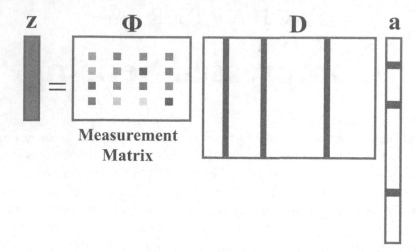

**Figure 4.1:** The linear compressive sensing system. **z** is the sensed signal, $\boldsymbol{\Phi}$ is the measurement matrix, **D** is the dictionary, and **a** is the sparse coefficient vector.

~$5S$ measurements are needed to recover an $S$-sparse wavelet coefficient vector. This translates to ~75% savings in a typical case [121].

## 4.1   MEASUREMENT MATRIX DESIGN

The encoding process in compressive sensing depends on the decoder used and vice-versa. Since we use linear measurements for encoding sparse signals, decoding or reconstruction can be performed using $\ell_0$ minimization or its convex version, the $\ell_1$ minimization

$$\min_{\mathbf{a}} \|\mathbf{a}\|_1 \text{ subj. to } \mathbf{z} = \boldsymbol{\Phi}\mathbf{D}\mathbf{a}. \tag{4.2}$$

We will analyze the properties and constructions of measurement matrices that will be suitable for $\ell_1$ minimization based compressive recovery.

### 4.1.1   THE RESTRICTED ISOMETRY PROPERTY

Let us denote the product dictionary $\boldsymbol{\Phi}\mathbf{D}$, which directly operates on the sparse vector **a**, as $\boldsymbol{\Psi}$. For the sake of analysis, let us assume that **D** is an identity matrix, which implies $\boldsymbol{\Psi} = \boldsymbol{\Phi}$ and $M = K$. In other words, the signals are assumed to be sparse in the canonical basis. In order to perform unique sparse recovery using $S$-sparse vectors whose coefficient locations are known, it is easy to see that all possible sub-matrices of $\boldsymbol{\Phi}$ of size $N \times S$ must be well-conditioned. This is expressed as

$$(1 - \delta) \leq \frac{\|\boldsymbol{\Phi}\mathbf{a}\|_2}{\|\mathbf{a}\|_2} \leq (1 + \delta) \tag{4.3}$$

for some $\delta > 0$. When the location of the non-zero entries are not known, the matrix $\boldsymbol{\Phi}$ must satisfy (4.3) for arbitrary $3K$-sparse vectors, and this condition is referred to as the restricted isometry property (RIP) [122]. Note that the sparse vector $\mathbf{a}$ has its energy concentrated in a few non-zero values, whereas the measurement vector $\boldsymbol{\Phi}\mathbf{a}$ has its energy spread more or less evenly over all its elements.

Although RIP is an intuitive property, verifying it for any matrix is computationally demanding, since, for $S$ non-zero coefficients, we need to check (4.3) for $\binom{K}{S}$ submatrices. However, matrices with entries obtained as i.i.d. realizations from certain distributions satisfy RIP with an overwhelming probability. One such distribution is the Gaussian distribution $\mathcal{N}(0, 1/N)$. The other distribution that is more amenable for computations is the Bernoulli distribution taking the value $1/\sqrt{N}$ and $-1/\sqrt{N}$ with a probability of 0.5. For measurement matrices obtained from these distributions, sparse and compressible signals of length $K$ can be recovered with high probability from only $N \geq cS \ln(K/S) \ll K$ measurements [119, 120]. Note that the analysis extends to the case where $\mathbf{D}$ is orthonormal and $\boldsymbol{\Phi}$ is obtained from an i.i.d. Gaussian distribution, the product matrix $\boldsymbol{\Psi}$ is also i.i.d. Gaussian, and, hence, in this case, $\boldsymbol{\Phi}$ is a universal measurement matrix. For a general matrix $\boldsymbol{\Psi}$, it is possible to characterize RIP using its mutual coherence $\mu$ [123].

In order to understand why random constructions of $\boldsymbol{\Phi}$ will work, we will look first at a classical result that claims that projecting arbitrary data onto random linear subspaces approximately preserves pairwise distances of data samples with high probability. The well-known Johnson-Lindenstrauss (JL) lemma [124] states that, for a set $\mathcal{Q}$ of $K-$dimensional points of size $|\mathcal{Q}|$, and a positive integer $N > N_0 = O(\ln(|\mathcal{Q}|)/\epsilon^2)$, the following property holds for $\epsilon \in (0, 1)$, using the random map $\boldsymbol{\Phi}$ obtained using the aforementioned distributions,

$$(1 - \epsilon)\|\mathbf{u} - \mathbf{v}\|_2^2 \leq \|\boldsymbol{\Phi}\mathbf{u} - \boldsymbol{\Phi}\mathbf{v}\|_2^2 \leq (1 + \epsilon)\|\mathbf{u} - \mathbf{v}\|_2^2, \tag{4.4}$$

for $\mathbf{u}, \mathbf{v} \in \mathcal{Q}$. Proving the JL lemma for random matrices consists of two steps. In the first step, assuming $\mathbf{D} = \mathbf{I}$, it is proved that the length of any $\mathbf{a} \in \mathbb{R}^K$ is preserved on average, after random projection,

$$\mathbf{E}(\|\boldsymbol{\Phi}\mathbf{a}\|_2^2) = \|\mathbf{a}\|_2^2. \tag{4.5}$$

In the second step, it is shown that the random variable $\|\boldsymbol{\Phi}\mathbf{a}\|_2^2$ is strongly concentrated about its expected value,

$$P(\big|\|\boldsymbol{\Phi}\mathbf{a}\|_2^2 - \|\mathbf{a}\|_2^2\big| \geq \epsilon\|\mathbf{a}\|_2^2) \leq 2e^{-Nc_0(\epsilon)}, 0 < \epsilon < 1, \tag{4.6}$$

taken over all possible random matrices of size $N \times K$, and $c_0(\epsilon)$ depends only on $\epsilon$ such that $c_0(\epsilon) > 0$. By applying (4.6), using a union bound to a set of differences between all points in $\mathcal{Q}$, we can prove the JL lemma [125]. One of the simplest proofs of RIP can be obtained directly using the concentration inequality (4.6) on embedding $S$-sparse vectors in random low-dimensional subspaces [125]. The proof considers only unit $\ell_2$ norm sparse vectors since the embedding using $\boldsymbol{\Phi}$ is a linear operation. The main strategy in the proof is to construct a covering of points in each $S-$dimensional subspace and apply (4.6) to the cover, using the union bound. The result is then extended to all possible $S-$dimensional signals lying in $\binom{K}{S}$ subspaces. This proof also makes it clear that the RIP condition (4.3) will hold with high probability for any random construction of $\boldsymbol{\Phi}$ that satisfies the concentration inequality (4.6). In addition to RIP, the deterministic sparsity threshold given in (2.15) can also be computed for a specific product dictionary $\boldsymbol{\Phi}$ to understand the behavior of compressive recovery. However, in practice, phase-transition characteristics of the measurement/coefficient system are more useful for understanding sparsity thresholds that guarantee recovery with high probability [63]. A brief description of phase-transition diagrams has already been provided in Section 2.3.1.

### 4.1.2   GEOMETRIC INTERPRETATION

Since compressive recovery is essentially an $\ell_1$ minimization problem, it can also be interpreted geometrically. Considering the recovery of the sparse solution $\mathbf{a}$ from $\mathbf{z} = \boldsymbol{\Phi}\mathbf{a}$, it is clear that the general solution lies in the translated null space $H(\boldsymbol{\Phi}) + \mathbf{a}$. This is illustrated in Figure 4.2 for a two-dimensional case. Note that we still assume $\mathbf{D} = \mathbf{I}$ and, hence, $\boldsymbol{\Phi} = \boldsymbol{\Psi}$. Since $\boldsymbol{\Phi}$ is random, the null space will also be randomly oriented. Figure 4.2 also compares the solutions obtained by blowing the $\ell_1$ and the $\ell_2$ norm balls until they touch the translated null space. As it can be seen, $H(\boldsymbol{\Phi}) + \mathbf{a}$ will touch the $\ell_1$ ball with high probability at its pointed vertex, leading to recovery of the correct sparse solution. In contrast, the $\ell_2$ penalty will find a dense solution in most of the cases. The seemingly small difference between the $\ell_1$ and the $\ell_2$ norms affects the solution substantially. It is clear from the above discussion that the null space $H(\boldsymbol{\Phi})$ is quite significant in the compressive recovery procedure. The necessary and sufficient conditions on the null space, for ensuring proper recovery of sparse and compressible signals, are formalized as the null space property (NSP) by the authors in [126].

### 4.1.3   OPTIMIZED MEASUREMENTS

According to the recovery condition for sparse representations given in (2.15), it can be seen that the sparsity threshold improves as the mutual coherence of $\boldsymbol{\Psi}$ decreases, i.e., as the columns of $\boldsymbol{\Psi}$ becomes more incoherent. Note that various mutual coherence measures of $\boldsymbol{\Psi}$ can be obtained by considering the off-diagonal elements of the Gram matrix $\boldsymbol{\Psi}^T\boldsymbol{\Psi}$. It is possible to design an optimized measurement matrix $\boldsymbol{\Phi}$ such that the columns of the product matrix $\boldsymbol{\Psi}$ are incoherent. In [127], an algorithm is proposed to obtain $\boldsymbol{\Phi}$ such that the $t-$averaged mutual coherence of $\boldsymbol{\Psi}$ is reduced. The $t-$averaged mutual coherence is the mean of the absolute value of the off-diagonal elements in the Gram matrix $\boldsymbol{\Psi}^T\boldsymbol{\Psi}$.

The measurement matrix can also be optimized such that the equivalent Gram matrix $\boldsymbol{\Psi}^T\boldsymbol{\Psi} \approx \mathbf{I}$ [28]. When compared to the previous approach, this leads to a more well-behaved

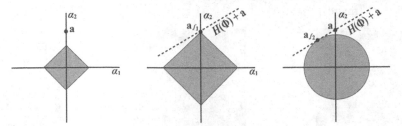

**Figure 4.2:** Recovering the sparse solution, from the random projections, using $\ell_1$ and $\ell_2$ minimization. Since the solution lies in the translated null space $H(\boldsymbol{\Phi}) + \mathbf{a}$, we can arbitrarily expand the corresponding norm balls until they touch the solution space, as illustrated in the middle ($\ell_1$) and right ($\ell_2$) figures.

algorithm. The optimized measurement matrix is now computed as $\boldsymbol{\Phi} = \boldsymbol{\Gamma}\mathbf{V}^T$, where $\mathbf{V}\boldsymbol{\Lambda}\mathbf{V}^T$ is the eigenvalue decomposition of $\mathbf{DD}^T$. Denoting the eigenvalues of $\mathbf{DD}^T$ to be $\{\lambda_1, ..., \lambda_M\}$ in descending order of magnitude, $\boldsymbol{\Gamma} = [\boldsymbol{\Gamma}_1\ \boldsymbol{\Gamma}_2]$, where $\boldsymbol{\Gamma}_1 \in \mathbb{R}^{N \times N}$ is a diagonal matrix with $\{\lambda_1^{-\frac{1}{2}}, \cdots, \lambda_N^{-\frac{1}{2}}\}$ as its diagonal elements, and $\boldsymbol{\Gamma}_2 \in \mathbb{R}^{N \times (M-N)}$ is a matrix of all zeros. Each column of the product matrix is given by $\boldsymbol{\psi}_j = \boldsymbol{\Phi}\mathbf{d}_j$, where $\mathbf{d}_j$ is the $j^{\text{th}}$ column of the dictionary $\mathbf{D}$.

## 4.2 COMPRESSIVE SENSING OF NATURAL IMAGES

We demonstrate the performance of compressed recovery in natural images using random as well as optimized measurement matrices with learned dictionaries. The images are divided into non-overlapping patches of size $8 \times 8$, and sensing and recovery is performed on a patch-by-patch basis. The noisy measurement process is be expressed as

$$\mathbf{z} = \boldsymbol{\Phi}\mathbf{Da} + \boldsymbol{\eta}, \tag{4.7}$$

where the dictionary $\mathbf{D}$ is learned from training data and $\boldsymbol{\eta}$ is the Gaussian noise vector added to the measurement process. In our example case, we consider global multilevel dictionaries, designed using randomly chosen patches from the BSDS dataset (Section 3.2.3). The entries in the random measurement matrices are independent realizations from $\mathcal{N}(0, 1)$, and the optimized measurement matrices are obtained using the method described in Section 4.1.3.

For the *Lena* image, we vary the desired number of measurements $N$ between 4 and 32 (6.25% and 50%) for a patch of size $8 \times 8$. In addition, we add Gaussian noise to the measurements such that the measurement SNR is at 15 dB. Recovery of each patch is performed by computing the sparse code, $\mathbf{a}$, for the measurements (random or optimized) using the product dictionary $\boldsymbol{\Psi} = \boldsymbol{\Phi}\mathbf{D}$. Finally, the full dimensional patch is reconstructed as $\mathbf{x} = \mathbf{Da}$. Figures 4.3 and 4.4 show the recovered images obtained with random and optimized measurement matrices respectively. In each case, the number of measurements ($N$) and the PSNR of the reconstructed image are

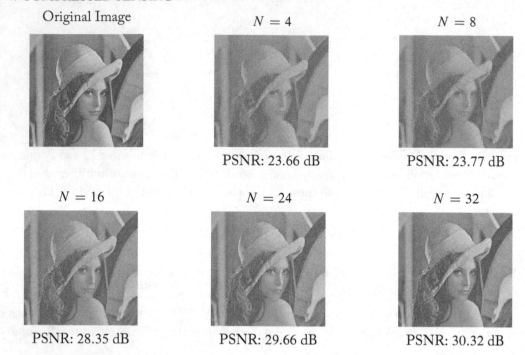

**Figure 4.3:** Compressive recovery of the *Lena* image using random measurements. The number of measurements is varied and the sparse codes are obtained using a global multilevel dictionary (Section 3.2.3). For each case, the reconstructed images and the corresponding peak signal-to-noise ratio (PSNR) are shown.

also shown. As it can be observed, increasing the number of measurements results in an improved recovery performance.

## 4.3   VIDEO COMPRESSIVE SENSING

In this section, we briefly discuss the compressive sensing of videos and related problems, such as video reconstruction, temporal super-resolution, and direct feature extraction for activity recognition. In contrast to the large breadth of work in the compressive sensing of images, video CS has only recently gained attention. In addition to the spatial priors that one uses to represent static images, video data calls for effective utilization of spatio-temporal priors. In the following, we discuss three aspects related to compressive sensing of videos: (a) frame-to-frame video reconstruction from traditional linear compressive measurements; (b) coupling of novel prior models and novel sensing strategies; and (c) direct feature extraction from compressed data for higher-level inference problems.

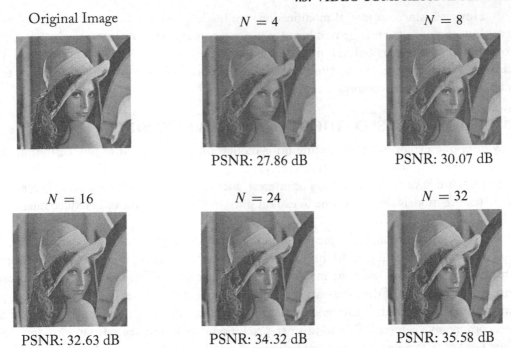

Original Image

$N = 4$
PSNR: 27.86 dB

$N = 8$
PSNR: 30.07 dB

$N = 16$
PSNR: 32.63 dB

$N = 24$
PSNR: 34.32 dB

$N = 32$
PSNR: 35.58 dB

**Figure 4.4:** Compressive recovery of the *Lena* image using measurements optimized to the sparsifying dictionary. The number of measurements is varied and the sparse codes are obtained using a global multilevel dictionary (Chapter 3). For each case, the reconstructed images and the corresponding peak signal-to-noise ratio (PSNR) are shown.

## 4.3.1 FRAME-BY-FRAME COMPRESSIVE RECOVERY

At the simplest level, one can design compressive sensing strategies, which work on a frame-by-frame manner, acquiring compressive measurements of each frame individually. Such approaches have been explored in [128, 129, etc.], where one considers a video as a sequence of static compressive measurements. A similar framework extends to hyper-spectral imagers, where one considers the "time-axis" to be replaced by the "wavelength-axis," where images in each wavelength are compressively sensed using the same measurement matrix [130]. Typical sparsifying priors that have been used for recovering the data-cube include spatial wavelet priors as in [130], or 3-D space-time wavelets [129].

At the recovery stage, one can exploit temporal correlations to reduce the complexity of frame-by-frame reconstruction as done in [131, 132]. In this class of approaches, it is assumed that the spatial support of video frames changes slowly, thereby allowing one to quickly obtain an approximate reconstruction using the support set of a previous time-instant, and updating the support set by a simpler reconstruction algorithm for the residual.

However, since no special modifications are made at the acquisition stage, many of these algorithms require a fairly large number of measurements at each time instant, which is generally proportional to the sparsity level of each individual frame. In the next section, we shall discuss examples where a more careful choice of spatio-temporal priors, with modification of sensing strategies, afford better compression properties.

## 4.3.2    MODEL-BASED VIDEO COMPRESSIVE SENSING

How does one exploit domain specific prior models on video signals to reduce the number of measurements required for reconstruction? While no general theory has been proposed for this problem, there have been a few cases of interest, such as assumptions of a periodic scene, or dynamic texture models that allow one to exploit prior models to significantly reduce measurement rates.

Veeraraghavan *et al.* [133] propose a compressive sensing framework of periodic scenes using coded strobing techniques. In this work, the goal was to image high-speed scenes by implicitly enhancing the temporal sampling rate of a conventional camera. This was achieved by measuring random dot-products of the time-varying signals arriving at each pixel individually, using spatio-temporal masks in front of the sensor array. This architecture was termed as a "programmable pixel compressive camera." Due to this per-pixel modulation, one records random dot products of the fast scene at a much lower sampling rate than otherwise necessary. For the special case of high-speed periodic scenes, such as a rotating mill tool, the authors proposed to use the standard Fourier basis as the sparsifying basis, since periodic signals are naturally sparse in the Fourier domain.

Linear dynamical systems (LDS) have proved to be a robust model to represent a wide variety of spatio-temporal data, including dynamic textures [134], traffic scenes [135], and human activities [136, 137]. Let $\{\mathbf{x}_t\}_{t=0}^{T}$ be a sequence of frames indexed by time $t$. The LDS model parameterizes the evolution of $\mathbf{x}_t$ as follows:

$$\mathbf{x}_t = \mathbf{C}\mathbf{s}_t + \mathbf{w}_t \quad \mathbf{w}_t \sim \mathcal{N}(\mathbf{0}, \mathbf{R}), \mathbf{R} \in \mathbb{R}^{M \times M} \tag{4.8}$$

$$\mathbf{s}_{t+1} = \mathbf{A}\mathbf{s}_t + \mathbf{v}_t \quad \mathbf{v}_t \sim \mathcal{N}(\mathbf{0}, \mathbf{Q}), \mathbf{Q} \in \mathbb{R}^{d \times d}, \tag{4.9}$$

where $\mathbf{s}_t \in \mathbb{R}^d$ is the hidden state vector, $\mathbf{A} \in \mathbb{R}^{d \times d}$ the transition matrix, and $\mathbf{C} \in \mathbb{R}^{M \times d}$ is the observation matrix.

In [138], an algorithm is derived to estimate the LDS parameters $(\mathbf{A}, \mathbf{C})$ from compressive measurements $\mathbf{z}_t = \boldsymbol{\Phi}\mathbf{x}_t$. A novel measurement strategy involving static and dynamic measurements is used to enable the estimation of parameters. The advantages of estimating LDS parameters over the original frames are easy to see. Firstly, the LDS parameters are much lower-dimensional than the total number of pixels in a given video. Further, both spatial sparsity and temporal correlations can be exploited using this model.

### 4.3.3   DIRECT FEATURE EXTRACTION FROM COMPRESSED VIDEOS

While reconstruction of videos from compressive measurements has seen growing interest, much less attention has been devoted to the question of whether higher-level inference tasks, such as detection and recognition of activities, can be performed without reconstructing the original video. Most higher-level video analysis tasks, such as object, activity recognition, and scene recognition, require non-linear feature-extraction techniques. Typical features useful for activity analysis include histogram of gradients (HOG) [139], optical flow [140], 3D SIFT [141], contours [136] etc. It is quite difficult to obtain such complex features directly from the compressive measurements without an intermediate step of signal reconstruction. However, an intermediate reconstruction step is often time-consuming, and in many cases not necessary. In this context, there is a growing need to explore novel features that retain robustness and accuracy, yet are amenable to extraction directly from compressed measurements.

An application of significant interest is human action recognition from surveillance cameras. However, human action recognition consists of several modules, such as moving human detection, tracking, and then action recognition, all of which are hard to perform in a compressive setting.

Background subtraction from compressively sensed data is a first step toward solving the moving-object detection problem. Recent work shows that tasks like background subtraction [142] are possible from compressive videos at far lower measurements than necessary for video reconstruction. In [142], it was shown that, for a far-field video and a static background, one can assume that moving objects in the scene are spatially sparse. If the goal is simply to perform background subtraction, then one can achieve high compression rates, yet be able to perform adaptive background differencing, leading to moving object detection. The tracking problem is much more involved and, so far, not much work exists that studies tracking directly from compressively sensed data.

For action recognition, a framework for non-linear dynamical invariant extraction from compressive videos was proposed in [143]. When a sequence of images is acquired by a compressive camera, the measurements are generated by a sensing strategy that maps the $M$−dimensional image space to an $N$−dimensional observation space. The overall mapping consists of a transformation $\mathbf{F}$ from the 3-D scene-space to image-space, with the addition of noise $\mathbf{n}$ in the sensor, followed by the measurement matrix $\boldsymbol{\Psi}$, which gives the measurements $\mathbf{z}_t$,

$$\mathbf{x}_t = \mathbf{F} \circ S_t + \mathbf{n}_t \qquad (4.10)$$
$$\mathbf{z}_t = \boldsymbol{\Phi}\mathbf{x}_t. \qquad (4.11)$$

Here, $S_t$ conceptually refers to a 3-D model of the scene with a human performing an action and $\mathbf{x}_t$ is the vectorized image obtained by transforming the 3-D scene to the image space. Assuming that the changes in the scene are due to a human performing an activity, we seek features that can be extracted directly from the sequence of measurements $\{\mathbf{z}_t\}$. Since we do not intend to reconstruct the image sequence, we are restricted in our ability to extract meaningful features. However, the JL lemma suggests that the general geometric relations of a set of points in a high-dimensional space can be preserved by certain embeddings into a low-dimensional space. In the case of compressive sensing, this embedding is achieved by projection onto the random linear subspace corresponding to the measurement matrix $\boldsymbol{\Phi}$. The preserving of relative distances between images under such an embedding allows one to explore the notion of recurrence plots [144] for dynamical invariant extraction.

Recurrence plots (RPs) are a visualization tool for dynamical systems. A recurrence matrix defined as

$$R(i, j) = \theta(\epsilon - \|\mathbf{x}_i - \mathbf{x}_j\|_2), \qquad (4.12)$$

where $\mathbf{x}_t$ is the observed time series and $\theta(.)$ is the Heaviside step function. RPs are shown to capture the system's behavior and be distinctive for different dynamical systems. At the time instant $t$, the compressive measurement of the image observation (the $t^{th}$ frame of the video sequence) is $\mathbf{z}_t \in \mathbb{R}^M$. Thus, if a sufficient number of measurements are taken, then with high probability the RPs for the compressed $\{\mathbf{z}_t\}$ and uncompressed signals $\{\mathbf{x}_t\}$ will be the same. This is a straightforward consequence of the JL lemma. On visualizing these recurrence matrices as images as shown in figure 4.5, it is clear that different activities give rise to widely different *recurrence textures*.

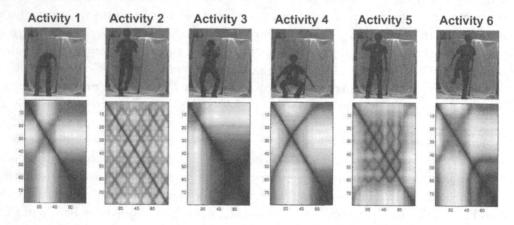

**Figure 4.5:** Row1: Examples of different activities from UMD dataset; Row2: Corresponding recurrence texture representations of the actions.

Motivated by this, [143] pose the problem of the classification of the dynamical system as a texture recognition problem. In table 4.1, we present average recognition results on the UMD activity dataset, when the compression ratio was varied across a broad range of values. We observe that the proposed framework works very well across a wide variety of compression factors. These are encouraging and positive results, which suggest that compressive cameras can afford carefully chosen higher-level inference without reconstruction. Furthermore, this method is also

**Table 4.1:** Reconstruction-free activity recognition rate for different compression factors on the UMD activity dataset. The recognition rates are quite stable even at very high compression rates

| Compression factor | Recognition Rate |
|---|---|
| Uncompressed | 90% |
| 100 | 90% |
| 400 | 86% |
| 800 | 84% |
| 1000 | 81% |
| 1200 | 80% |

**Table 4.2:** Reconstruction-free classification of traffic patterns at different compression ratios (in %) on the UCSD Traffic Dataset [135]

| Compression ratio | Expt.1 | Expt.2 | Expt.3 | Expt.4 |
|---|---|---|---|---|
| 25× | 92.06 | 92.19 | 85.94 | 92.06 |
| 150× | 88.89 | 78.13 | 78.13 | 82.54 |
| 300× | 87.30 | 82.81 | 76.56 | 82.54 |

applicable to the classification of dynamic textures, such as traffic scenes. Results of traffic-pattern classification from compressive videos are shown in Table 4.2 as a function of compression rate.

# CHAPTER 5

# Sparse Models in Recognition

Sparse learning, in addition to providing interpretable and meaningful models, is efficient for large-scale learning. Though the paradigm of sparse coding has been very effective in modeling natural signals, its ability to discriminate different classes of data is not inherent. Since sparse coding algorithms aim to reduce only the reconstruction error, they do not explicitly consider the correlation between the codes, which is crucial for classification tasks. As a result, the codes obtained for similar signals or features (extracted from the signals) may be quite different in traditional sparse coding. In addition, the overcomplete nature of the dictionary makes the sparse coding process highly sensitive to the variance in the features. Hence, adapting this representative model to perform discriminative tasks requires the incorporation of supervisory information into the sparse coding and dictionary learning problems. By introducing the prior knowledge on the sparsity of signals into the traditional machine learning algorithms, novel discriminative frameworks can be developed.

## 5.1    A SIMPLE CLASSIFICATION SETUP

We begin by presenting a basic classification system that exploits the structured sparsity of signals in a class to learn sparse templates for signal classification. In template matching, a prototype of the training signals is typically generated and can be compared with the test signal to be recognized, taking into account the transformations that might have occurred in the test signal. The metric for similarity is often a correlation measure, but, if the template is modeled statistically, a likelihood measure can be used. In the following formulation, we generate class-specific statistical templates using sparse codes of the training data.

Let us consider an oracle source model in which the sources correspond to the multiple classes of data, and each class uses a subset of elements chosen from an unknown generating dictionary, $\mathbf{D}^G$. It is assumed that the observation vectors are generated by linearly combining a fixed set of basis functions with random weights that belong to a Gaussian distribution. For any class $p$, this can be expressed as

$$\mathbf{x}_p = \sum_{k \in \Lambda_p} a_{p,k} \mathbf{d}_k^G + \mathbf{n}, \tag{5.1}$$

where $\mathbf{x}_p$ is an observation vector, $a_{p,k}$ are the random weights, $\mathbf{d}_k^G$ are the elements chosen from the generating dictionary $\mathbf{D}^G$ (with $K$ elements), and $\mathbf{n}$ is Additive White Gaussian Noise (AWGN). The coefficient vector for class $p$, $\mathbf{a}_p \sim N(\mathbf{m}_p, \Sigma_p)$ is the random vector with $a_{p,k}$ as its elements. For simplicity, it is assumed that the random weights are independent and, hence, $\Sigma_p$ is a diagonal matrix. The set $\Lambda_p$ contains the indices of a subset of dictionary atoms from $\mathbf{D}^G$, that were used in the generation of $\mathbf{x}_p$. Therefore, $\Lambda_p$ contains the indices of the elements in $\mathbf{a}_p$ that have a non-zero variance. It is important to note that we are dealing with sparsely representable signals and, hence, typically $|\Lambda_p| \ll K$. An illustration of the source model for the case with 3 sources is given in Figure 5.1.

In the setup considered, there are totally $P$ classes and each class contains $T$ training observations, $\mathbf{X}_p = \{\mathbf{x}_{p,1}, ..., \mathbf{x}_{p,T}\}$. In this demonstrative formulation, we have assumed an ideal scenario where all signals in a class share a similar set of patterns from an unknown dictionary, combined with randomly generated weights. Though this is far from reality, we use this example to demonstrate how this prior knowledge can be efficiently employed to build a classification system.

An estimation of the unknown dictionary $\mathbf{D}^G$, by exploiting the structure sparsity of signals in a class, and the computation of the coefficient matrix $\mathbf{A}_p = [\mathbf{a}_{p,1}...\mathbf{a}_{p,T}]$ for each $\mathbf{X}_p$, can be performed using an iterative procedure comprising the following steps:

- *Representation*: Using all labeled data samples in a class, we can perform group sparse coding to exploit the fact that all samples in the class share a common non-zero support in their coefficient vectors. In other words, we approximate all the vectors in a single class, $\mathbf{X}_p$, using different linear combinations of the same set of dictionary elements. As a result, the maximum number of dictionary atoms that can be used in the approximation for each source needs to be fixed. Finally, the coefficient matrices of all classes are stacked together to obtain $A = [A_1...A_P]$.

- *Dictionary Update*: The unknown dictionary can be updated by generalizing the update step in the K-SVD algorithm, described in Chapter 3, to consider multiple training vectors for each source. For the dictionary update step, the optimization problem can be posed as

$$\min_{\mathbf{D}} \left\| \mathbf{X}^{(1)} - \mathbf{D}\mathbf{A}^{(1)} \right\|_F^2 + ... + \left\| \mathbf{X}^{(T)} - \mathbf{D}\mathbf{A}^{(T)} \right\|_F^2 \qquad (5.2)$$

$$= \min_{\mathbf{D}} \left\| \mathbf{E}_k^{(1)} - \mathbf{d}_k a_T^{k(1)} \right\|_F^2 + ... + \left\| \mathbf{E}_k^{(T)} - \mathbf{d}_k a_T^{k(T)} \right\|_F^2$$

$$= \min_{\mathbf{D}} \left\| \left[ \mathbf{E}_k^{(1)} ... \mathbf{E}_k^{(T)} \right] - \mathbf{d}_k \left[ a_T^{k(1)} ... a_T^{k(T)} \right] \right\|_F^2$$

$$= \min_{\mathbf{D}} \left\| \mathbf{E}_k^{tot} - \mathbf{d}_k \mathbf{a}_T^{k(tot)} \right\|_F^2 . \qquad (5.3)$$

In effect, the updated dictionary atom $\mathbf{d}_k$ can be determined using a rank-1 approximation of the matrix $\mathbf{E}_k^{tot}$, which is carried out using singular value decomposition.

As described earlier, each training vector is an independent observation and each column of $\mathbf{A}_p$, $\mathbf{a}_{p,t}$ is a realization of the random vector $\mathbf{a}_p$. Since it is assumed that $\mathbf{a}_p$ follows a Gaussian

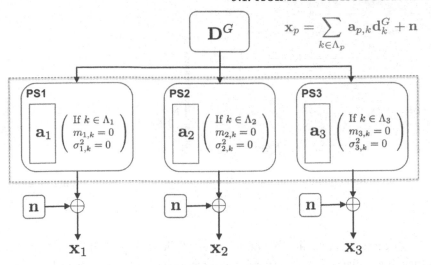

**Figure 5.1:** The oracle source model for generating the signals. PS1, PS2 and PS3 are the sources and each source represents a class that uses a fixed set of dictionary atoms from the generating dictionary. For simplicity, it is assumed that the coefficient values in each class are randomly generated and the observations are possibly corrupted by additive noise.

distribution, we can estimate its mean and variance as follows:

$$\hat{\mathbf{m}}_p = \frac{1}{T} \sum_{t=1}^{T} \mathbf{a}_{p,t}, \tag{5.4}$$

$$\hat{}_p = \frac{1}{T-1} \sum_{t=1}^{T} (\mathbf{a}_{p,t} - \hat{\mathbf{m}}_p)(\mathbf{a}_{p,t} - \hat{\mathbf{m}}_p)^T, \tag{5.5}$$

where $\hat{\mathbf{m}}_p$ is the mean estimate and $\hat{}_p$ is the estimate of covariance for the oracle source $p$, parameterized by the statistical template, $\tau_p = \{\mathbf{m}_p, {}_p\}$. Since the covariance matrix is diagonal, we denote the vector containing the diagonal elements in ${}_p$ by $\hat{}_p^d = diag(\hat{}_p)$. Hence, there are $P$ sets of mean and covariance matrices that characterize each of the probability sources.

Given a test data sample, $\mathbf{z}$, the classification problem is to identify the probability source that could have generated it, in the Maximum Likelihood (ML) sense. Let $\mathbf{w}$ be the coefficient vector of $\mathbf{z}$ using the estimated dictionary ${}_G$. Hence, the estimate of its class is

$$\hat{p} = \operatorname*{argmax}_p L(\mathbf{w}|\hat{\tau}_p) = \operatorname*{argmax}_p \sum_{k=1}^{K} \left( -\frac{1}{2}\ln(2\sigma\,\hat{}_{p,k}^2) - \frac{1}{2}\frac{(w_k - \hat{m}_{p,k})^2}{\hat{}_{p,k}^2} \right), \tag{5.6}$$

where $\hat{m}_{p,k}$, $\hat{}_{p,k}^2$ and $w_k$ are the elements of the vectors $\hat{m}_p$, $\hat{}_p^d$ and $w$ respectively.

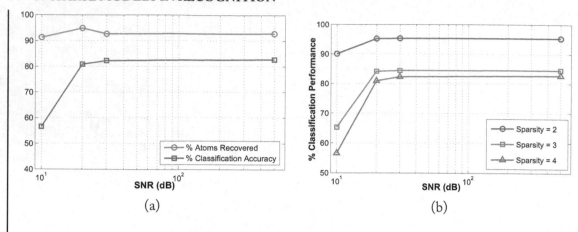

**Figure 5.2:** (a) Simulation results for the recovery of dictionary atoms and classification using the sparse coefficient templates when the sparsity of signals in each class is fixed at $S = 4$. (b) Performance of the classifier for different values of sparsity $S$ evaluated in the presence of varying levels of additive noise.

For a synthetic experiment, let us fix the dimension of the training vectors at 16 and the number of dictionary atoms to be 64. The synthetic signals are generated using the oracle source model in (5.1), with the additive noise adjusted according to the target SNR. The dictionary $\mathbf{D}^G$ consists of Gaussian random vectors with zero mean and unit variance. We fix the number of classes at $P = 1{,}500$, with $T = 25$ training observations in each class. For a particular class, the observations are generated by linearly combining $S$ randomly picked dictionary elements. Note that we allowed two different classes to share more than one dictionary atom. However, we merged classes with the exact same support. The same sparsity is used as the constraint while adapting the dictionary and computing the atomic representation. The weights of the linear combination are realized from independent Gaussian distributions with different non-zero means and unit variance.

We infer the unknown dictionary using the iterative procedure described earlier, and generate the statistical templates. For testing the classification, we generate 1,500 test observations from randomly picked sources and corrupt them with AWGN. The ratio of the number of recovered atoms to the actual number of atoms in $\mathbf{D}^G$ is shown in Figure 5.2(a) for $S = 4$. By exploiting the group sparsity of all signals in a class, we are able to estimate the underlying dictionary accurately, even under noisy observation conditions. On the other hand, the classification capability of this framework highly relies on the inherent sparsity of the signals in the different classes. Figure 5.2(b) illustrates the classification accuracy as a function of the target SNR, for different values of sparsity. It can be observed that the classifier performance improves when the signals are highly sparse with respect to $\mathbf{D}^G$. This simple formulation illustrates the fact that the discrimination power of the sparse codes can be improved by incorporating additional supervisory knowledge on the training examples.

## 5.2    DISCRIMINATIVE DICTIONARY LEARNING

Sparse representations using predefined and learned dictionaries have been extremely successful in the representation of images. However, in order for them to be successful in classification, explicit constraints for discrimination need to be incorporated while adapting the dictionary. It is well known that the traditional linear discriminant analysis (LDA) approach can aid in supervised classification by projecting the data onto a lower dimensional space where the inter-class separation is maximized while the intra-class separation is minimized [145]. If the sparse representation is constrained such that a discriminatory effect can be introduced in the coefficient vectors of different classes, the computed representation can be used in classification or recognition tasks.

Assume that the sparse codes of the $T$ training vectors $\{\mathbf{x}_i\}_{i=1}^T$ are given by $\{\mathbf{a}_i\}_{i=1}^T$ and the indices of the training vectors belonging to class $p$ are given by $\mathcal{C}_p$. Each training vector belongs to one of the $P$ classes. Denoting the mean and variance for each class as $\mathbf{m}_i$ and $\sigma_i^2$ respectively, they can be computed as the sample mean and variance estimates of coefficient vectors corresponding to a class. The Fisher's discriminant is defined as

$$F(\mathbf{A}) = \frac{\| \sum_{p=1}^P T_p (\mathbf{m}_p - \mathbf{m})(\mathbf{m}_p - \mathbf{m})^T \|_F^2}{\sum_{p=1}^P \sigma_p^2}, \tag{5.7}$$

where $\mathbf{m}$ is the sample mean estimate of all coefficient vectors and $T_p$ is the number of training vectors that belong to class $p$. Incorporating the Fisher's discriminant, the objective function that focuses only on discrimination can be written as

$$\operatorname*{argmax}_{\mathbf{A}} F(\mathbf{A}) - \lambda \sum_{i=1}^T \|\mathbf{a}_i\|_0, \tag{5.8}$$

whereas the objective function that combines both representation on a dictionary $\mathbf{D}$ and discrimination can be posed as

$$\operatorname*{argmax}_{\mathbf{A}} F(\mathbf{A}) - \lambda_1 \sum_{i=1}^T \|\mathbf{x}_i - \mathbf{D}\mathbf{a}_i\|_2^2 - \lambda_2 \sum_{i=1}^T \|\mathbf{a}_i\|_0. \tag{5.9}$$

This was one of the first models that included explicit discrimination constraints in sparse representation-based classification [146]. Models that perform supervised dictionary learning using supervised sparse coding [147], by directly incorporating the SVM model parameters [36], have also been proposed.

## 5.3 SPARSE-CODING-BASED SUBSPACE IDENTIFICATION

The sparse representation of signals, though not developed for classification tasks, can be discriminative, since it selects the most compact set of dictionary elements. Sparse representations can be effectively used for classification, if the overcomplete dictionary is constructed using the training samples [37, 148]. If a sufficient number of samples are available from each class, it is possible to represent a test sample using a small subset of training samples from the same class. Loosely speaking, computing the sparsest representation automatically discriminates between the different classes. It can clearly be observed that this approach is a generalization of the nearest-neighbor and the nearest-subspace algorithms.

The important challenge in performing feature-based recognition of objects and faces is in deciding a suitable low-dimensional image-level feature that is informative and still discriminative between classes. Several approaches, such as the Eigenfaces, the Fisherfaces, and the Laplacianfaces have been very successfully used to efficiently project high-dimensional data to low-dimensional feature spaces. However, the theory of compressed sensing suggests that even random projections can be employed and, hence, the choice of the suitable feature space in no longer critical in sparse representation frameworks. Formally stating the problem, we are provided with a set of face images from $k$ different classes and the images are vectorized and stacked as columns in the input matrix $\mathbf{X} = [\mathbf{X}_1, \mathbf{X}_2, ..., \mathbf{X}_P]$. Here, the matrix $\mathbf{X}_p = [\mathbf{x}_{p,1}, \mathbf{x}_{p,2}, ..., \mathbf{x}_{p,T_p}]$ contains the vectorized images from class $p$.

Several models have been proposed for exploiting the structure of $\mathbf{X}$ for recognition. An effective model for face recognition is to model samples from a single class to be lying on a linear subspace. Face images with varying lighting and expression have been shown to lie in a low-dimensional subspace, referred to as the *face subspace*. Given a sufficient number of images in each class, a test sample $\mathbf{z}$ from class $p$ will approximately lie in the linear span of the training samples from the same class.

$$\mathbf{z} = \sum_{i=1}^{T_p} \mathbf{a}_{p,i} \mathbf{x}_{p,i}. \tag{5.10}$$

However, the underlying class of the test sample is unknown to begin with and, hence, we evaluate a linear representation using all images in $\mathbf{X}$. The approximation can be obtained by picking the nearest neighbors to the test sample and computing the coefficients by solving a least squares problem. However, it has been found in [37] that using sparse coding is superior to using nearest-neighbor-based approaches. This is because, when there are a large number of samples in each class, the resulting coefficient vector is naturally sparse. In fact, the more sparse the coefficient vector $\mathbf{a}$, the easier it is to identify the class membership of $\mathbf{z}$. For the test sample, we compute the coefficient vector and evaluate the residual error with respect to every class. For the $p^{\text{th}}$ class we compute the residual error as

$$\mathbf{r}_p(\mathbf{z}) = \|\mathbf{z} - \mathbf{X}\delta_p(\mathbf{a})\|, \tag{5.11}$$

(a) Semi-supervised learning.

(b) Self-taught transfer learning.

**Figure 5.3:** Machine learning formalisms to include unlabeled data in supervised learning. The supervised task is to classify lamps and cameras and the self-taught learning framework uses randomly chosen unlabeled data.

where $\delta_p(\mathbf{a})$ is a new vector with only non-zero coefficients from $\mathbf{a}$ corresponding to the class $p$. Finally, the test sample is assigned to the class with the least residual error.

## 5.4 USING UNLABELED DATA IN SUPERVISED LEARNING

Supervised classification in machine learning often requires labeled data that are expensive and difficult to obtain in practice. However, the abundance of unlabeled data motivates the design of frameworks that can exploit this information, to perform supervised tasks. Self-taught transfer learning is a recent machine learning formalism based on sparse coding that provides a principled approach to the use of unlabeled data. Though semi-supervised learning allows unlabeled data, it makes an additional assumption that the labels of the data are just unobserved and can be labeled with the same labels as the supervised task. Figure 5.3 illustrates the two formalisms to include unlabeled data in supervised classification.

Self-taught learning is motivated by the observation that local regions of natural images demonstrate redundancy. In other words, even random images downloaded from the Internet will contain elementary features that are similar to those in the images we want to classify. To formally state the problem, we are provided with a labeled set of training examples $\{(\mathbf{x}_l^1, y^1), (\mathbf{x}_l^2, y^2), (\mathbf{x}_l^3, y^3), ...., (\mathbf{x}_l^{T_l}, y^{T_l})\}$, where $\mathbf{x}_l^i$ denotes the $i^{\text{th}}$ training example and $y^i$ is its corresponding label. In addition, we are provided with a set of unlabeled examples, $\mathbf{x}_u^1, \mathbf{x}_u^2, ..., \mathbf{x}_u^{T_u}$. No assumptions are made regarding the underlying distribution of the unlabeled examples. However, it is assumed that the examples are of the same type as the labeled examples, e.g., images, audio. The algorithm proposed in [38] uses the raw pixel intensities in $\{\mathbf{x}_u^i\}_{i=1}^{T_u}$ to learn the elementary features that comprise an image. As a result, it learns to represent images in terms of the features rather than in terms of the raw pixel intensities. These learned features are used to obtain an abstract or higher-level representation for the labeled data $\{\mathbf{x}_l^i\}_{i=1}^{T_l}$ as well.

The representative features are learned using a modified version of the sparse coding algorithm proposed by Olshausen and Field, which modeled the processing in the primary visual cortex of the humans. Given the set of unlabeled data,

$$\min_{\mathbf{D},\mathbf{a}} \sum_i \|\mathbf{x}_u^i - \sum_j a_j^i \mathbf{d}_j\|_2^2 + \beta \|\mathbf{a}^i\|_1 \qquad \text{subj. to} \qquad \|\mathbf{d}_j\|_2 \leq 1, \forall j. \qquad (5.12)$$

In other words, the algorithm learns an overcomplete dictionary using the patches from the unlabeled images. This optimization problem, though not convex jointly, is convex over the subset of variables $\mathbf{D}$ and $\mathbf{a} = \{\mathbf{a}^1, \ldots, \mathbf{a}^{Tu}\}$. The optimization over the coefficients is an $\ell_1$ regularized least squares problem, while the optimization over the dictionary is an $\ell_2$ constrained least squares problem. This iterative optimization can be performed using the dictionary learning approaches described in Chapter 3.

In general, it is often easy to obtain large amounts of unlabeled data from the Internet that share important features with the labeled data of interest. Hence, the dictionary learned using the unlabeled data can generalize well to represent data in the classification task. For each training input from the classification task, we evaluate sparse codes with $\mathbf{D}$, using error-constrained $\ell_1$ minimization. Finally, these features can be used to train standard supervised classifiers such as the SVM.

## 5.5  GENERALIZING SPATIAL PYRAMIDS

In complex visual recognition tasks, it is common to extract relevant local or global image descriptors instead of using the raw image pixels directly. Scene understanding is typically carried out by building a Bag of Words (BoW) model that collectively represents the set of local features in an image. However, the spatial information about the features is ignored when considering the orderless BoW model. The spatial information can also be exploited by aggregating the features (using histograms) at multiple spatial scales, and this process is referred to as constructing a spatial pyramid. Several state-of-the-art object recognition systems involve the construction of spatial pyramid features to achieve improved recognition performance.

The algorithm proposed in [40] partitions the image into increasingly finer regions and evaluates features in the local regions. Typically, local features such as the Scale Invariant Feature Transform (SIFT) or the Histogram of Oriented Gradients (HOG) descriptors are evaluated for small image patches, and the descriptors are coded using vector quantization. Since the local features demonstrate redundancy across multiple images, the descriptors can be coded efficiently using codebooks generated by standard procedures, such as the K-means clustering. In a given spatial scale, all code vectors in each sub-region can be aggregated by building histograms. As we move from a coarser to a finer spatial scale, the number of sub-regions increase. The aggregated feature provides a trade-off between translation invariance and spatial locality. In the coarsest spatial scale where the whole image is considered, the max pooling feature achieves translation invariance of the local patterns. On the other hand, as we proceed toward finer spatial scales,

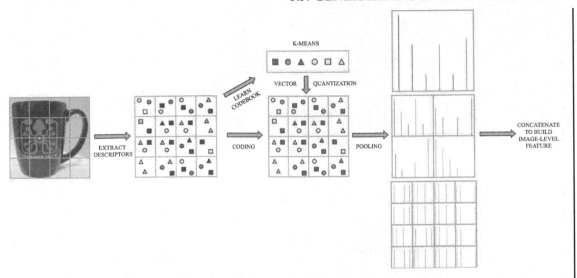

**Figure 5.4:** Illustration of the process of constructing the spatial pyramid feature for an image. Local descriptors obtained from the patches in the image are used to learn the codebook. Using the codebook, we perform vector quantization of the local descriptors to exploit the redundancy. Finally, we construct the bag-of-features model at multiple spatial scales. The resulting spatial pyramid feature is the concatenation of histograms from all scales.

spatial information is efficiently captured. By concatenating the set of histograms obtained at all spatial scales, we can construct the spatial pyramid. Figure 5.4 illustrates the steps involved in the spatial pyramid matching algorithm.

The spatial pyramid feature, when combined with a non-linear classifier, has been shown to achieve high recognition accuracies. Though, this approach has been very effective, the complexity of using a non-linear classifier is quite high. Using linear kernels with histogram features often leads to substantially worse results, partially due to the high quantization error in vector quantization. However, when using non-linear kernels, the computational cost of computing the kernel ($\mathcal{O}(T^3)$) and storage requirements ($\mathcal{O}(T^2)$) are quite high, particularly for large values of $T$ (total number of training features) [149]. Furthermore, as the number of support vectors scales to the number of training samples, testing cost also increases. As a result, there is a need to employ a different feature-extraction mechanism such that the classification can be performed efficiently using only linear classifiers.

The ScSPM algorithm proposed in [149] addresses this problem by computing sparse codes of the local image descriptors, and performing non-linear aggregation (or pooling). Given a pre-defined dictionary $\mathbf{D}$ and the set of descriptors $\mathbf{X}$, we can obtain the sparse codes using any algorithm described in Chapter 2. A suitable image feature can be constructed by applying a pre-defined aggregation function to the coefficient matrix $\mathbf{A}$. In the coefficient matrix, each row

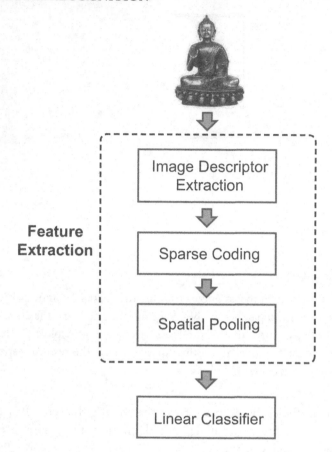

**Figure 5.5:** Demonstration of linear spatial pyramid matching based on sparse coding. Note that the vector quantization step in the conventional non-linear SPM is replaced by sparse coding, and the underlying spatial pooling process is max pooling.

corresponds to the response of all descriptors to a particular dictionary atom. Though a number of pooling functions can be used, the max pooling function has been found to perform the best. By performing max pooling on $\mathbf{A}$, we can obtain the feature vector $\mathbf{z} \in \mathbb{R}^K$ as follows:

$$z_k = \max\{|a_{k,1}|, |a_{k,2}|, \ldots, |a_{k,T}|\}, \text{ for } k = 1 \text{ to } K. \tag{5.13}$$

Here $|a_{k,i}|$ denotes the element in the $k^{\text{th}}$ row, $i^{\text{th}}$ column of the coefficient matrix; $T$ and $K$ denote the total number of descriptors in the image and dictionary atoms respectively. The max pooling process has been commonly employed in mathematical models for the visual cortex (V1) and empirically justified by many algorithms. Similar to the construction of the spatial pyramid feature, max pooling is performed across multiple locations, and different spatial scales of the image. Hence, the aggregated feature is more robust to local transformations when compared to using histograms. Figure 5.5 illustrates the steps involved in the ScSPM algorithm. Though the conventional spatial pyramid feature leads to very poor performances with a linear kernel, a linear SPM kernel based on sparse coding statistics can achieve high classification accuracy. This behavior can be attributed to the better approximation power of sparse coding in natural images and the discrimination power of the aggregated spatial pyramid feature.

## 5.5.1 SUPERVISED DICTIONARY OPTIMIZATION

In applications such as face recognition, evaluating a sparse representation of a test sample in terms of its training examples, and allowing some error for occlusion, fails to effectively handle variations such as pose and alignment. As a result, extracting image-level features that are invariant to global misalignment can result in improved recognition performance [150]. As described earlier, the ScSPM feature provides a trade-off between translation invariance and spatial locality. The efficiency of these features can be improved further by optimizing the dictionary $\mathbf{D}$ to obtain ScSPM features with higher discrimination.

For a training image $\mathbf{X}_t$, we can obtain the coefficient matrix $\mathbf{A}_t$ by performing sparse coding using a known dictionary $\mathbf{D}$. In order to construct the ScSPM feature, we perform max pooling of the coefficients in all cells of each spatial scale. We denote the set-level max pooled feature in the $c^{\text{th}}$ spatial cell of the $s^{\text{th}}$ spatial scale by $\mathbf{z}_c^s$. Assuming that there are $R$ spatial scales, in each scale the image is divided into $2^{s-1} \times 2^{s-1}$ non-overlapping cells. The hierarchical ScSPM feature is constructed by concatenating all set-level features.

$$\mathbf{z}_t = \eta_{max}(\mathbf{A}_t) = \bigcup_{s=1}^{R} \left[ \bigcup_{c=1}^{2^{s-1}} \mathbf{z}_c^s \right]. \tag{5.14}$$

For classification, let us assume a predictive model $f(\mathbf{z}_t, \mathbf{w})$, a loss function $\ell(y_t, f(\mathbf{z}_t, \mathbf{w}))$, where $y_k$ denotes the class label and $\mathbf{w}$ indicates the predictive model parameters. The objective function for obtaining the ScSPM features, dictionary, and the predictive model parameters can be expressed as

$$E(\mathbf{D}, \mathbf{w}, \{\mathbf{X}_t\}_{t=1}^{T}) = \sum_{t=1}^{T} \ell(y_t, f(\mathbf{z}_t, \mathbf{w})) + \lambda \|\mathbf{w}\|_2^2. \tag{5.15}$$

This optimization can be performed alternatively with respect to $\mathbf{D}$ and $\mathbf{w}$. Given the dictionary, solving for $\mathbf{w}$ is equivalent to training a classifier, and, hence, any standard algorithm can be used. However, in order to optimize $E$ over $\mathbf{D}$, we compute the gradient as

$$\frac{\partial E}{\partial \mathbf{D}} = \sum_{t=1}^{T} \frac{\partial \ell}{\partial \mathbf{D}} = \sum_{t=1}^{T} \frac{\partial \ell}{\partial f} \frac{\partial f}{\partial \mathbf{z}_t} \frac{\partial \mathbf{z}_t}{\partial \mathbf{A}_t} \frac{\partial \mathbf{A}_t}{\partial \mathbf{D}}. \tag{5.16}$$

This gradient is evaluated using implicit differentiation [150], and, finally, the dictionary is updated iteratively such that the different classes are separable.

## 5.6   LOCALITY IN SPARSE MODELS

In non-linear function learning problems, it is typical to overfit the data, particularly when the dimensionality of the data is much higher in comparison to the number of data samples. However, in several real problems we do not observe this so-called curse of dimensionality, since they often lie on a manifold with much smaller intrinsic dimensionality. Performing machine learning tasks using such data is to effectively exploit the intrinsic geometric properties of the manifold of interest. In recent years, the knowledge of the geometric properties has been utilized to devise sophisticated algorithms for tasks such as classification and clustering. One of the approaches that has been popular in the machine learning community is the use of non-linear dimensionality reduction (NLDR) techniques [151]. Their main aim is to identify a low-dimensional manifold onto which the high-dimensional data can be projected while preserving most of the local and/or global structure. Approaches such as Locally Linear Embedding, Hessian LLE, and Laplacian Eigenmaps preserve the local information of the manifold. In contrary, global methods such as Isomap, semidefinite embedding, etc. try to preserve both global and local relationships. However, this class of techniques assumes that the data is embedded in a high-dimensional ambient vector space, i.e., the manifold is an embedded sub-manifold of some vector space. If such an embedding can be found, one can apply a variety of methods developed for vector-spaces in conjunction with one of several NLDR techniques.

### 5.6.1   LOCAL SPARSE CODING

Locally Linear Embedding (LLE) is an unsupervised learning algorithm that exploits the fact that the local geometry of a non-linear function can be well approximated using a linear model. When the dictionary $\mathbf{D}$ represents the set of anchor points that characterize the local geometry, the idea of using sparse coding to model the local neighborhood of data merits attention. However, sparse coding, in the absence of additional constraints, tries to reduce the error of the representation without any consideration on locality. It is possible to include additional locality constraints by considering the general class of the weighted $\ell_1$ minimization problems,

$$\min_{\mathbf{a}} \sum_{k=1}^{K} w(k)|a_k| \quad \text{subject to} \quad \|\mathbf{x} - \mathbf{Da}\|_2 \leq \epsilon, \tag{5.17}$$

where $w(1), ..., w(K)$ are positive weights. It can clearly be seen that large weights could be used to encourage zero entries in the sparse code $\mathbf{a}$. The LCC algorithm proposed in [39] computes local sparse codes using weights based on the Euclidean distance between the data sample to be coded and the dictionary atoms, as given by

$$w(k) = \|\mathbf{x} - \mathbf{d}_k\|_2^2, \tag{5.18}$$

where $\mathbf{d}_k$ is the $k^{\text{th}}$ column of the dictionary matrix $\mathbf{D}$. An alternate distance function has been proposed in [152],

$$w(k) = \|\mathbf{x} - (\mathbf{x}^T \mathbf{d}_k)\mathbf{d}_k\|_2^2. \tag{5.19}$$

This weighting scheme directly considers the coherence between the normalized dictionary elements and the data sample $\mathbf{x}$.

Since the weighted $\ell_1$ minimization in (5.17) is computationally expensive, an approximate method for locality constrained linear coding was proposed in [153]. The LLC algorithm employs the following criteria:

$$\min_{\mathbf{A}} \sum_{i=1}^{T} \|\mathbf{x}_i - \mathbf{D}\mathbf{a}_i\|_2^2 + \lambda \|\mathbf{w}_i \odot \mathbf{a}_i\|_2^2 \text{ subj. to } \mathbf{1}^T \mathbf{a}_i = 1, \forall i, \tag{5.20}$$

where $\odot$ denotes element-wise multiplication and $\mathbf{w}_i$ measures the similarity between the data sample and all the dictionary atoms. The distance metric used is

$$w_i(k) = \exp\left(\frac{\|\mathbf{x}_i - \mathbf{d}_k\|_2}{\sigma}\right), \forall k, \tag{5.21}$$

where $\sigma$ is used to adjust the control of the magnitude of weights in the neighborhood of a data sample. In order to speed up this procedure, the $N$-nearest dictionary atoms are first identified and a smaller linear system is solved using a least squares procedure on the chosen dictionary atoms. This reduces the computational complexity from $\mathcal{O}(K^2)$ to $\mathcal{O}(K + N^2)$, where $K$ denotes the number of dictionary atoms and $N \ll K$.

## 5.6.2 DICTIONARY DESIGN

The joint optimization problem for local sparse coding and dictionary learning can be expressed as

$$\left\{\hat{\mathbf{D}}, \{\hat{\mathbf{a}}_i\}_{i=1}^{T}\right\} = \min_{\mathbf{D}, \{\mathbf{a}_i\}_{i=1}^{T}} \sum_{i=1}^{T} \|\mathbf{x}_i - \mathbf{D}\mathbf{a}_i\|_2^2 + \|\mathbf{W}_i \mathbf{a}_i\|_1, \tag{5.22}$$

additionally constraining $\|\mathbf{d}_k\|_2 = 1, \forall k = \{1, \ldots, K\}$. In order to solve for local sparse codes, we rewrite the weighted minimization problem in (5.22) as

$$\hat{\mathbf{a}}_i = \operatorname*{argmin}_{\mathbf{a}_i} \|\mathbf{x}_i - \mathbf{D}\mathbf{W}_i^{-1}\mathbf{a}_i\|_2^2 + \lambda \|\mathbf{a}_i\|_1, \text{ for } i = 1 \text{ to } T. \tag{5.23}$$

Here $\mathbf{W}_i$ is a diagonal matrix containing the weights corresponding to the dictionary elements in $\mathbf{D}$ computed as

$$\mathbf{W}_i(k, k) = \|\mathbf{x}_i - \mathbf{d}_k\|_2^2, \text{ for } k = 1 \text{ to } K. \tag{5.24}$$

Setting $\boldsymbol{\Gamma}_i = \mathbf{D}\mathbf{W}_i^{-1}$, (5.23) can be expressed as

$$\hat{\mathbf{a}}_i = \operatorname*{argmin}_{\mathbf{a}_i} \|\mathbf{x}_i - \boldsymbol{\Gamma}_i \mathbf{a}_i\|_2^2 + \lambda \|\mathbf{a}_i\|_1, \text{ for } i = 1 \text{ to } T. \tag{5.25}$$

This is equivalent to standard sparse coding with the modified dictionary $\boldsymbol{\Gamma}_i$. Adapting a dictionary for local sparse coding can be performed using the approach described in [152]. The neighborhood relationship between the dictionary and the data samples is preserved by fixing the coefficients during the dictionary update. Given the sparse codes and the weighting matrix, the dictionary update step can be expressed as

$$\hat{\mathbf{D}} = \min_{\mathbf{D}} \sum_{i=1}^{T} \|\mathbf{x}_i - \mathbf{D}\mathbf{z}_i\|_2, \text{ subj. to } \|\mathbf{d}_k\|_2^2 = 1, \forall k. \tag{5.26}$$

Here $\mathbf{z}_i = \mathbf{W}_i^{-1}\mathbf{a}_i$ denotes the reweighted coefficient vector. As shown in [152], the dictionary atom $\mathbf{d}_k$ can be updated as

$$\hat{\mathbf{d}}_k = \frac{\mathbf{E}_k \mathbf{z}_{k,r}^T}{\|\mathbf{E}_k \mathbf{z}_{k,r}^T\|_2}. \tag{5.27}$$

It can be observed that this dictionary update is equivalent to computing the weighted mean of the residual error vectors, $\mathbf{E}_k$, where the weights are obtained using local sparse coding. By optimizing the dictionary for the local sparse codes and aggregating the codes using the ScSPM procedure, improved recognition performance can be obtained.

## 5.7   INCORPORATING GRAPH EMBEDDING CONSTRAINTS

The inherent low dimensionality of data usually is exploited in order to achieve improved performances in various computer vision- and pattern-recognition tasks. Several supervised, semi-supervised, and unsupervised machine learning schemes can be unified under the general framework of graph embedding. Locality preserving projections (LPP) is a graph embedding approach that computes projection directions, such that the pairwise distances of the projected training samples in the neighborhood are preserved. Defining the data to be $\mathbf{X} = [\mathbf{x}_1 \ldots \mathbf{x}_T] \in \mathbb{R}^{M \times T}$, we can construct an undirected graph $G$ with the training samples as vertices, and the similarity between the neighboring training samples is coded in the affinity matrix $\mathbf{W} \in \mathbb{R}^{T \times T}$. Let us define the graph laplacian as $\mathbf{L} = \mathbf{B} - \mathbf{W}$, where $\mathbf{B}$ is a degree matrix with each diagonal element containing the sum of the corresponding row or column of $\mathbf{L}$. The $d$ projection directions for LPP, $\mathbf{V} \in \mathbb{R}^{M \times d}$, can be computed as

$$\hat{\mathbf{V}} = \underset{\text{trace}(\mathbf{V}^T \mathbf{XBX}^T \mathbf{V})=\mathbf{I}}{\operatorname{argmin}} \text{trace}(\mathbf{V}^T \mathbf{XLX}^T \mathbf{V}). \tag{5.28}$$

The LPP formulation does not incorporate supervisory label information while computing the embedding, which is crucial in classification tasks. Hence, we can also consider the version of graph embedding as applied to supervised and semi-supervised learning problems. Local discriminant embedding (LDE) [154] incorporates supervised label information of training data in the graph $G$, and defines another graph $G'$ that considers the inter-class relationships. Denoting the graph laplacian of $G'$ by $\mathbf{L}'$, the optimal directions for linear embedding in LDE can be obtained by solving

$$\hat{\mathbf{V}} = \underset{\text{trace}(\mathbf{V}^T \mathbf{XL}' \mathbf{X}^T \mathbf{V})=\mathbf{I}}{\text{argmin}} \text{trace}(\mathbf{V}^T \mathbf{XLX}^T \mathbf{V}). \tag{5.29}$$

Furthermore, when only a subset of the training data is labeled, a semi-supervised graph embedding approach, referred to as semi-supervised discriminant analysis (SDA), has been developed [155]. Incorporating these graph embedding principles into sparse representation-based learning schemes will allow the design of novel recognition frameworks.

## 5.7.1 LAPLACIAN SPARSE CODING

Sparse coding aims to obtain a parsimonious representation for the data using the basis functions in a given dictionary. The effectiveness of sparse coding in recognition is affected by two main limitations: (a) Due to the overcomplete nature of the dictionary, even a small variance in the local features might result in significantly different responses to the dictionary atoms; and (b) the relationship between the local features is not considered in the coding process. In order to better characterize the relationship between the local features, and make the coding process robust, the underlying graph structure of the data samples can be exploited. Unlike coding similar features together using group coding, explicit regularization terms can be added to the sparse coding problem to preserve the consistency of the sparse codes for similar features. Assuming that the set of features $\mathbf{X}$ can be sparsely represented using a dictionary, $\mathbf{D} \in \mathbb{R}^{M \times K}$, the authors in [156] proposed the *Laplacian Sparse Coding* approach to obtain sparse codes, $\mathbf{A} \in \mathbb{R}^{K \times T}$, that preserve the neighborhood structure of $G$ by solving

$$\{\hat{\mathbf{D}}, \hat{\mathbf{A}}\} = \underset{\mathbf{D}, \mathbf{A}}{\text{argmin}} \|\mathbf{X} - \mathbf{DA}\|_F^2 + \gamma \sum_i \|\mathbf{a}_i\|_1 + \beta \text{trace}(\mathbf{ALA}^T)$$

$$\text{subj. to } \forall j, \|\mathbf{d}_j\|_2 \leq 1, \tag{5.30}$$

where $\mathbf{a}_i$ is the sparse code for the sample $\mathbf{x}_i$, and $\mathbf{d}_j$ is the $j^{\text{th}}$ column of $\mathbf{D}$. As observed in the previous dictionary learning formulations, the problem in (5.30) is not jointly convex and, hence, alternatively solved to obtain the dictionary and the laplacian sparse codes. Instead of learning an embedding explicitly as in (5.28), here we impose the graph structure on the resulting sparse codes.

## 5.7.2 LOCAL DISCRIMINANT SPARSE CODING

In this section we present sparse coding formulations, for supervised and semi-supervised learning respectively, that employ regularization terms for exploiting the graph structure. The local discriminant sparse coding (LDSC) and semi-supervised discriminant sparse coding (SDSC) algorithms, proposed in [157], build the graphs $G$ and $G'$, using the methods adopted in LDE and SDA respectively. Figure 5.6 illustrates the two sparse coding approaches.

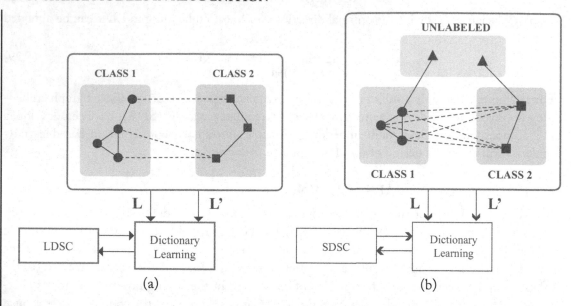

(a)                                    (b)

**Figure 5.6:** Proposed supervised and semi-supervised learning approaches: (a) Local Discriminant Sparse Coding; (b) Semi-Supervised Discriminant Sparse Coding. Note that solid and dotted lines denote the edges in the graphs $G$ and $G'$ respectively. In (b), $G'$ is fully connected within a class also (the dotted lines for $G'$ overlap with the solid lines).

The proposed optimization problem for dictionary learning and sparse coding with graph embedding constraints can be expressed as

$$\{\hat{\mathbf{D}}, \hat{\mathbf{A}}\} = \underset{\mathbf{D}, \mathbf{A}}{\text{argmin}} \, \|\mathbf{X} - \mathbf{D}\mathbf{A}\|_F^2 + \gamma \sum_i \|\mathbf{a}_i\|_1 + \beta \text{trace}(\mathbf{A}\mathbf{L}\mathbf{A}^T) \tag{5.31}$$

$$\text{subj. to trace}(\mathbf{A}\mathbf{L}'\mathbf{A}^T) = c, \forall j \, \|\mathbf{d}_j\|_2 \le 1, \tag{5.32}$$

where $\gamma$, $\beta$, and $c$ are positive constants. The codes generated are sparse, reconstruct the data well, and preserve the neighborhood structure of the graphs $G$ and $G'$. Sparse codes for the novel test samples can be computed using the learned dictionary $\hat{\mathbf{D}}$. Note that (5.31) is a general formulation and subsumes (5.30) as a special case.

When labeled training data is available, the problem in (5.31) can be solved to obtain dictionaries that produce highly discriminative sparse codes, by using the graph laplacians as defined in LDE. Let the label for the training data $\mathbf{x}_i$ be $y_i$ and let $\mathcal{N}_k(i)$ denote the set of $k$-nearest samples of $\mathbf{x}_i$. For LDSC, the entries in the affinity matrix $\mathbf{W}$ and $\mathbf{W}'$ are defined as [154]

$$w_{ij} = \begin{cases} 1 & \text{if } y_i = y_j \text{ AND } [i \in \mathcal{N}_k(j) \text{ OR } j \in \mathcal{N}_k(i)], \\ 0 & \text{otherwise}, \end{cases} \tag{5.33}$$

$$w'_{ij} = \begin{cases} 1 & \text{if } y_i \neq y_j \text{ AND } [i \in \mathcal{N}_{k'}(j) \text{ OR } j \in \mathcal{N}_{k'}(i)], \\ 0 & \text{otherwise.} \end{cases} \quad (5.34)$$

The graph laplacians $\mathbf{L}$ and $\mathbf{L}'$, used in (5.31), are then constructed using $\mathbf{W}$ and $\mathbf{W}'$.

When the training set consists of both unlabeled data, $\mathbf{X}_u \in \mathbb{R}^{M \times T_u}$, and labeled data, $\mathbf{X}_\ell \in \mathbb{R}^{M \times T_\ell}$, such that $\mathbf{X} = [\mathbf{X}_\ell \ \mathbf{X}_u]$, the dictionary $\mathbf{D}$ and sparse codes can be obtained using the graph laplacians defined by SDA. The SDSC algorithm is a semi-supervised approach that augments the labeled training data using unlabeled training samples in the neighborhood. The entries in the affinity matrices are given by [155]

$$w_{ij} = \begin{cases} 1/T_{y_i} + \alpha s_{ij} & \text{if } y_i = y_j, \\ \alpha s_{ij} & \text{otherwise,} \end{cases} \quad (5.35)$$

$$w'_{ij} = \begin{cases} 1/T_\ell & \text{if both } \mathbf{x}_i \text{ and } \mathbf{x}_j \text{ are labeled,} \\ 0 & \text{otherwise,} \end{cases} \quad (5.36)$$

where

$$s_{ij} = \begin{cases} 1 & \text{if } i \in \mathcal{N}_k(j) \text{ OR } j \in \mathcal{N}_k(i), \\ 0 & \text{otherwise.} \end{cases} \quad (5.37)$$

Here $T_{y_i}$ is the number of samples in the class specified by $y_i$ and $\alpha$ is a parameter that adjusts the relative importance of labeled and unlabeled data.

The problem in (5.31) is solved as an alternating minimization procedure that optimizes the dictionary while fixing the sparse codes, and vice-versa. When sparse codes are fixed, dictionary learning is a convex problem and, hence, can efficiently be solved using various approaches. When the dictionary is fixed, sparse code for any data sample $\mathbf{x}_i$ can be obtained by solving

$$\hat{\mathbf{a}}_i = \underset{\mathbf{a}_i}{\arg\min} \ \|\mathbf{x}_i - \mathbf{D}\mathbf{a}_i\|_2^2 + \gamma\|\mathbf{a}_i\|_1 + \beta(l_{ii}\|\mathbf{a}_i\|_2^2 + 2\mathbf{a}_i^T\mathbf{A}_{\backslash i}\mathbf{l}_{\backslash i})$$
$$\text{subj. to } l'_{ii}\|\mathbf{a}_i\|_2^2 + 2\mathbf{a}_i^T\mathbf{A}_{\backslash i}\mathbf{l}'_{\backslash i} = c/T, \quad (5.38)$$

where $l_{ii}$ and $l'_{ii}$ are the $i^{th}$ diagonal elements of $\mathbf{L}$ and $\mathbf{L}'$, $\mathbf{A}_{\backslash i}$ is the coefficient matrix with its $i^{th}$ column removed, and $\mathbf{l}_{\backslash i}$, $\mathbf{l}'_{\backslash i}$ are the $i^{th}$ columns of $\mathbf{L}$, $\mathbf{L}'$ with their $i^{th}$ entries removed. The objective is convex but non-differentiable where the coefficient value is zero, and we also have a non-convex (quadratic equality) constraint. Hence, a modified version of sequential quadratic programming can be used to solve for the codes [157].

## 5.8 KERNEL METHODS IN SPARSE CODING

Though the linear generative model of sparse coding has been effective in several image-understanding problems, exploiting the non-linear similarities between the training samples can result in more efficient models. It is typical in machine learning methods to employ the *Kernel Trick* to learn linear models in a feature space that captures the non-linear similarities. The Kernel Trick maps the non-linear separable features into a feature space $\mathcal{F}$ using a transformation $\Phi(.)$, in which similar features are grouped together. The transformation $\Phi(.)$ is chosen such that $\mathcal{F}$ is a Hilbert space with the reproducing kernel $\mathcal{K}(.,.)$ and, hence, the non-linear similarity between two samples in $\mathcal{F}$ can be measured as $\mathcal{K}(\mathbf{x}_i, \mathbf{x}_j) = \Phi(\mathbf{x}_i)^T \Phi(\mathbf{x}_j)$. Note that the feature space is usually high-dimensional (sometimes infinite) and the closed-form expression for the transformation $\Phi(.)$ may be intractable or unknown. Therefore, we simplify the computations by expressing them in terms of inner products $\Phi(\mathbf{x}_i)^T \Phi(\mathbf{x}_j)$, which can be replaced using $\mathcal{K}(\mathbf{x}_i, \mathbf{x}_j)$, whose value is always known. This is referred to as the Kernel Trick. Note that, in order for a kernel to be valid, the kernel function or the kernel matrix should be symmetric positive semidefinite according to Mercer's theorem [158].

## 5.8.1 KERNEL SPARSE REPRESENTATIONS

Computing sparse representations for data samples can be performed efficiently in a high-dimensional feature space using the kernel trick [159]. Let us define a feature-mapping function $\Phi : \mathbb{R}^M \mapsto \mathbf{R}^G$, where $M < G$, that maps the data samples and the dictionary atoms to a high-dimensional feature space. We denote the data sample $\mathbf{x}$ in the feature space as $\Phi(\mathbf{x})$ and the dictionary by $\Phi(\mathbf{D}) = [\Phi(\mathbf{d}_1), \Phi(\mathbf{d}_2), ..., \Phi(\mathbf{d}_K)]$. The kernel similarities $\mathcal{K}(\mathbf{x}_i, \mathbf{x}_j) = \Phi(\mathbf{x}_i)^T \Phi(\mathbf{x}_j)$, $\mathcal{K}(\mathbf{d}_k, \mathbf{x}) = \Phi(\mathbf{d}_k)^T \Phi(\mathbf{x})$, and $\mathcal{K}(\mathbf{d}_k, \mathbf{d}_l) = \Phi(\mathbf{d}_k)^T \Phi(\mathbf{d}_l)$ can be computed using pre-defined kernel functions, such as the Radial Basis Function (RBF) or the polynomial kernel. All further computations in the feature space should be performed exclusively using kernel similarities. The problem of sparse coding in can be posed in the feature space as

$$\min_{\mathbf{a}} \|\Phi(\mathbf{x}) - \Phi(\mathbf{D})\mathbf{a}\|_2^2 + \lambda \|\mathbf{a}\|_1. \tag{5.39}$$

Expanding the objective in (5.39) we obtain

$$\Phi(\mathbf{x})^T \Phi(\mathbf{x}) - 2\mathbf{a}^T \Phi(\mathbf{D})^T \Phi(\mathbf{x}) + \mathbf{a}^T \Phi(\mathbf{D})^T \Phi(\mathbf{D})\mathbf{a} + \lambda \|\mathbf{a}\|_1,$$
$$= \mathbf{K}_{\mathbf{xx}} - 2\mathbf{a}^T \mathbf{K}_{\mathbf{Dx}} + \mathbf{a}^T \mathbf{K}_{\mathbf{DD}}\mathbf{a} + \lambda \|\mathbf{a}\|_1, \tag{5.40}$$
$$= F(\mathbf{a}) + \lambda \|\mathbf{a}\|_1. \tag{5.41}$$

Here, $\mathbf{K}_{\mathbf{xx}}$ is the element $\mathcal{K}(\mathbf{x}, \mathbf{x})$, $\mathbf{K}_{\mathbf{Dx}}$ is a $K \times 1$ vector containing the elements $\mathcal{K}(\mathbf{d}_k, \mathbf{x})$, for $k = \{1, ..., K\}$, and $\mathbf{K}_{\mathbf{DD}}$ is a $K \times K$ matrix containing the kernel similarities between the dictionary atoms. As it can be easily observed, the modified objective function is similar to the sparse coding problem, except for the use of the kernel similarities. Hence, the kernel sparse coding problem can be solved using the feature-sign search algorithm or LARS efficiently. However, it is important to note that the computation of kernel matrices incurs additional complexity. Since the dictionary is fixed in (5.41), $\mathbf{K}_{\mathbf{DD}}$ is computed only once and the complexity of computing $\mathbf{K}_{\mathbf{Dx}}$ grows as $O(MK)$. Updating the atoms of $\mathbf{D}$, using the standard dictionary learning procedures, is not straightforward and, hence, we can employ a fixed-point procedure for dictionary update [159].

## 5.8.2   KERNEL DICTIONARIES IN REPRESENTATION AND DISCRIMINATION

Optimization of dictionaries in the feature space can be carried out by reposing the dictionary learning procedures using only the kernel similarities. Such non-linear dictionaries can be effective in yielding compact representations, when compared to approaches such as the kernel PCA, and in modeling the non-linearity present in the training samples. In this section, we will present the formulation of a kernel dictionary learning procedure, and demonstrate its effectiveness in representation and discrimination. For simplicity, we build the kernel version of the K-lines clustering algorithm, and illustrate how a linear procedure can be efficiently reposed using the Kernel Trick. The K-lines clustering procedure (Chapter 3) iteratively performs a least squares fit of $K$ 1-D subspaces to the training data. Since sparse coding aims to approximate a given data sample using a union of subspaces, designing dictionaries using the cluster centroids from K-lines clustering can be effective. Note that the K-SVD algorithm is a generalization of this clustering procedure. Hence, the kernel K-SVD algorithm proposed in [160] can also be generalized from the kernel K-lines procedure described here.

In order to perform K-lines clustering in the feature space, let us compute the cluster centroids using a linear and possibly iterative mechanism, instead of using an SVD. The objective function for K-lines clustering can be rewritten as

$$\{\mathbf{d}_k, \{a_{k,i}\}\} = \operatorname*{argmin}_{\mathbf{d}_k, \{a_{k,i}\}} \sum_{i \in \mathcal{C}_k} \|\mathbf{x}_i - a_{k,i}\mathbf{d}_k\|_2^2 \text{ subj. to } \|\mathbf{d}_k\|_2^2 = 1, \qquad (5.42)$$

where $\{a_{k,i}\}$ is the set of coefficients for the data samples with $i \in \mathcal{C}_k$. Note that (5.42) can be solved using an alternating minimization procedure, by fixing one of $\mathbf{d}_k$ or $\{a_{k,i}\}$ and solving for the other. Fixing $\mathbf{d}_k$, we can compute $\{a_{k,i}\}$ as $a_{k,i} = \mathbf{x}_i^T \mathbf{d}_k, \forall i \in \mathcal{C}_k$. Incorporating the constraint $\|\mathbf{d}\|_2 = 1$ and assuming $\{a_{k,i}\}$ to be known, we can compute the cluster center as

$$\mathbf{d}_k = \sum_{i \in \mathcal{C}_k} a_{k,i}\mathbf{x}_i \Big/ \|\sum_{i \in \mathcal{C}_k} a_{k,i}\mathbf{x}_i\|_2. \qquad (5.43)$$

This update is equivalent to computing the normalized weighted-mean of samples in a cluster. The centroid and the coefficients can be obtained for each cluster by iterating between coefficient computation and (5.43) for a sufficient number of times.

In order to rewrite this formulation using matrices, let us denote the association of training vectors to cluster centers by a membership matrix $\mathbf{Z} \in \mathbb{R}^{T \times K}$, where $z_{ik} = 1$ if and only if $i \in \mathcal{C}_k$. Hence, the cluster assignment problem can be solved by computing $\mathbf{A} = \mathbf{X}^T\mathbf{D}$, and then setting $\mathbf{Z} = g(\mathbf{A})$, where $g(.)$ is a function that operates on a matrix and returns 1 at the location of absolute maximum of each row and zero elsewhere. We define the matrix

$$\mathbf{H} = \mathbf{Z} \odot \mathbf{A}, \tag{5.44}$$

where $\odot$ indicates the Hadamard product. The centroid update can be performed as

$$\mathbf{D} = \mathbf{X}\mathbf{H}\boldsymbol{\Gamma}, \tag{5.45}$$

which is the matrix version of the update in (5.43). Here, $\boldsymbol{\Gamma}$ is a diagonal matrix with the $\ell_2$ norm of each column in $\mathbf{X}\mathbf{H}$. This ensures that the columns of $\mathbf{D}$ are normalized.

The kernel version of this clustering procedure can be developed by solving the coefficient computation and centroid update using kernel similarities. Let $\Phi(\mathbf{X})$ and $\Phi(D)$ respectively denote the matrix of all feature vectors and cluster centroids in the kernel domain. The kernel matrix $\mathbf{K_{XX}} \in \mathbb{R}^{T \times T}$ is a Gram matrix of all data samples. The coefficient matrix, $\mathbf{A}$, in the feature space can be computed as

$$\mathbf{A} = \Phi(\mathbf{X})^T \Phi(\mathbf{D}), \tag{5.46}$$

and the membership matrix in the feature space is given by $\mathbf{Z} = g(\mathbf{A})$. Hence, the cluster centers in the feature space can be computed as

$$\Phi(\mathbf{D}) = \Phi(\mathbf{X})\mathbf{H}\boldsymbol{\Gamma}, \tag{5.47}$$

where $\mathbf{H} = \mathbf{Z} \odot \mathbf{A}$. The normalization term is computed as

$$\boldsymbol{\Gamma} = diag((\Phi(\mathbf{X})\mathbf{H})^T \Phi(\mathbf{X})\mathbf{H})^{1/2} = diag(\mathbf{H}^T \mathbf{K_{XX}} \mathbf{H})^{1/2}, \tag{5.48}$$

where $diag(.)$ is an operator that returns a diagonal matrix with the diagonal elements being the same as that of the argument matrix. Combining (5.46), (5.47) and (5.48), we obtain

$$\mathbf{A} = \Phi(\mathbf{X})^T \Phi(\mathbf{X})\mathbf{H}\boldsymbol{\Gamma} = \mathbf{K_{XX}}\mathbf{H}\boldsymbol{\Gamma}. \tag{5.49}$$

Hence, kernel K-lines clustering can be performed by updating $\mathbf{A}$ and $\mathbf{H}$ in every iteration and the membership matrix $\mathbf{Z}$ once every $L$ iterations.

- *Representation*: Kernel sparse coding can be used as an alternative to approaches such as kernel PCA for efficient data representation. Though complete reconstruction of the underlying data from the kernel sparse codes requires computation of pre-images [], novel test samples can be well approximated using the learned kernel dictionaries. As a demonstration, we consider the class of digit 2 from the USPS dataset [161] and use a subset of images for training a kernel dictionary using kernel K-lines clustering. We then compute the sparse code for a novel test sample $\mathbf{z}$, different from the training set, and compute the reconstruction error as $\|\Phi(\mathbf{z}) - \Phi(\mathbf{D})\mathbf{a}\|_2^2$. Figure 5.7 shows the reconstruction error obtained for a test sample for different values of sparsity, $\{1, \ldots, 20\}$.

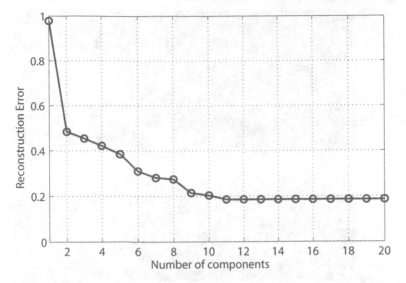

**Figure 5.7:** Reconstruction error for a novel test sample using kernel sparse coding, for different values of sparsity.

- *Discrimination*: In addition to efficiently modeling data samples, kernel sparse coding is well suited for supervised learning tasks. Since the non-linear similarities between the training samples are considered while learning the dictionary, the resulting codes are highly discriminative. For an example demonstration, we consider 100 training samples each from 3 different classes in the USPS dataset (Digits 3, 4 and 7). We obtain the kernel sparse codes for all the samples and compute the normalized cross correlation between the sparse features. In cases of high discrimination, we expect the features belonging to a class to be highly similar to each other, compared to samples from other classes. The block-wise structure in the normalized correlation plot in Figure 5.8 evidences the discrimination power of the kernel sparse codes.

### 5.8.3    COMBINING DIVERSE FEATURES

The use of multiple features to characterize images more precisely has been a very successful approach for complex visual recognition tasks. Though this method provides the flexibility of choosing features to describe different aspects of the underlying data, the resulting representations are high-dimensional and the descriptors can be very diverse. Hence, there is a need to transform these features to a unified space, that facilitates the recognition tasks, and construct low-dimensional compact representations for the images in the unified space.

Let us assume that a set of $R$ diverse descriptors is extracted from a given image. Since the kernel similarities can be used to fuse the multiple descriptors, we need to build the base

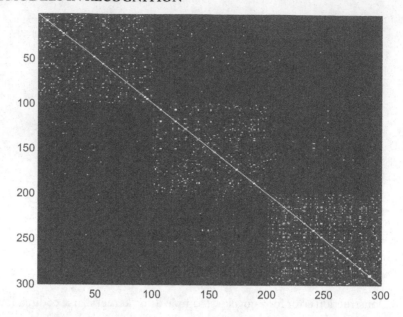

**Figure 5.8:** Similarity between the kernel sparse codes of samples drawn from 3 different classes in the USPS dataset. Since the kernel codes of samples belonging to the same class are highly similar, we observe a block-wise structure in the normalized correlation plot.

kernel matrix for each descriptor. Given a suitable distance function $d_r$, that measures the distance between two samples, for the feature $r$, we can construct the kernel matrix as

$$\mathbf{K}_r(i, j) = \mathcal{K}_r(\mathbf{x}_i, \mathbf{x}_j) = \exp(-\gamma d_r^2(\mathbf{x}_i, \mathbf{x}_j)), \tag{5.50}$$

where $\gamma$ is a positive constant. Computing this kernel function is convenient, since several image descriptors and corresponding distance functions have been proposed in the literature. Given the $R$ base kernel matrices, $\{\mathbf{K}_r\}_{r=1}^R$, we can construct the ensemble kernel matrix as

$$\mathbf{K} = \sum_{r=1}^R \beta_r \mathbf{K}_r, \quad \forall \beta_r \geq 0. \tag{5.51}$$

Note that this is not the only way to construct the ensemble matrix. For example, the descriptors can be alternatively fused as

$$\mathbf{K} = \mathbf{K}_1 \odot \mathbf{K}_2 \odot \ldots \odot \mathbf{K}_R, \tag{5.52}$$

where $\odot$ denotes the Hadamard product between two matrices. Performing sparse coding using the ensemble kernel matrices will result in highly efficient codes for recognition.

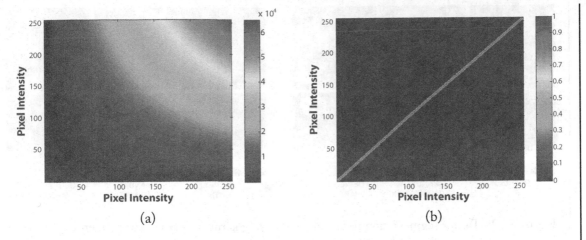

**Figure 5.9:** Similarity between grayscale pixel intensities (0 to 255). (a) Linear similarity ($y_i y_j$); and (b) non-linear similarity ($\mathcal{K}(y_i, y_j)$) using an RBF kernel.

## 5.8.4 APPLICATION: TUMOR IDENTIFICATION

In this section, we describe the application of kernel sparse codes to automatically segment active tumor components from T1-weighted contrast-enhanced Magnetic Resonance Images (MRI). Sparse coding algorithms typically are employed for vectorized patches in the images, or for descriptors. However, in this pixel-based tumor segmentation problem, we need to obtain sparse codes for representing each pixel in the image. This is trivial if we use the traditional sparse coding approach, since $M = 1$ in this case. Furthermore, in order to discriminate between the pixels belonging to multiple segments, we may need to consider the non-linear similarity between them. Hence, we can resort to using kernel sparse coding to obtain codes for each pixel [162].

Assume that the $T$ pixels in an image are represented by $\mathbf{X} = [x_i]_{i=1}^T$. In order to make a pixel-wise decision if it belongs to an active tumor region, we build the intensity kernel using the Radial Basis Function (RBF) kernel of the form $\mathcal{K}_I(x_i, x_j) = \exp(-\gamma(x_i - x_j)^2)$. Since the intensities of the tumor pixels are very similar, the intensity kernel can lead to highly discriminative sparse codes. The difference between the linear similarity of grayscale pixel intensities (0 to 255) and the non-linear similarities obtained using the RBF kernel ($\gamma = 0.3$) are illustrated in Figure 5.9 (a) and (b), respectively. The linear similarities depend predominantly on the individual intensities of the pixels and not on the closeness of intensities. When the RBF kernel is used, the pixel intensities that are close to each other have high non-linear similarity irrespective of the intensities. Pixels with intensities that are far apart have zero non-linear similarity. Therefore, the pixelwise sparse codes that we obtain using such a kernel will also follow a similar behavior, i.e., pixels with intensities close to each other will have similar sparse codes.

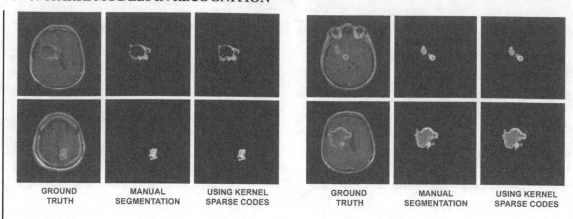

| GROUND TRUTH | MANUAL SEGMENTATION | USING KERNEL SPARSE CODES | | GROUND TRUTH | MANUAL SEGMENTATION | USING KERNEL SPARSE CODES |

**Figure 5.10:** Tumor segmentation results obtained using kernel sparse coding with ensemble kernel matrices. Both intensity and locality kernels are used to determine if each pixel belongs to an active tumor region.

Furthermore, in order to localize the tumor regions in the image, we need to incorporate additional constraints to ensure connectedness among pixels in a segment. This can be addressed by building a spatial locality kernel and fusing it with the intensity kernel. The spatial locality kernel $\mathbf{K}_L$ is constructed as

$$\mathbf{K}_L(i, j) = \mathcal{K}_L(x_i, x_j) = \begin{cases} \exp^{\|\mathbf{L}_i - \mathbf{L}_j\|_2^2}, & \text{if } j \in \mathcal{N}(i), \\ 0, & \text{otherwise.} \end{cases} \tag{5.53}$$

Here, $\mathcal{N}(i)$ denotes the neighborhood of the pixel $x_i$, and $\mathbf{L}_i$ and $\mathbf{L}_j$ are the locations of the pixels, $x_i$ and $x_j$, respectively. We combine the intensity and spatial locality kernel matrices as $\mathbf{K} = \mathbf{K}_I \odot \mathbf{K}_L$. Note that, when combining kernel matrices, we need to ensure that the resulting kernel matrix also satisfies the Mercer's conditions.

The sparse codes obtained with a dictionary learned in the ensemble feature space model the similarities with respect to both intensity and location of pixels. A set of training images, with active tumor regions, is used to learn a kernel dictionary with the kernel K-lines clustering procedure. It is assumed that the locations of the tumor pixels are known in the ground truth training images. Using the kernel sparse codes belonging to tumor and non-tumor regions, we learn 2-class linear SVM to classify the regions. For a test image, we obtain the required ensemble kernel matrices and compute the kernel sparse codes using the learned dictionary. Finally, the SVM classifier can be used to identify the pixels belonging to an active tumor region. Figure 5.10 shows the original MRI and the active tumor region as identified by this algorithm. For reference, a manual segmentation result performed by an expert is also shown. The impact of combining diverse features using kernel sparse coding is evidenced by the accurate segmentation results.

# Bibliography

[1] D. Martin, C. Fowlkes, D. Tal, and J. Malik, "A database of human segmented natural images and its application to evaluating segmentation algorithms and measuring ecological statistics," in *Proceedings of the 8th International Conference on Computer Vision*, July 2001, vol. 2, pp. 416–423. DOI: 10.1109/ICCV.2001.937655. 2, 36

[2] D.J. Field, "Relations between the statistics of natural images and the response properties of cortical cells," *Journal of the Optical Society of America*, vol. 4, pp. 2379–2394, 1987. DOI: 10.1364/JOSAA.4.002379. 2

[3] E. Simoncelli, "Modeling the joint statistics of images in the wavelet domain," *Proceedings of the SPIE 44th Annual Meeting*, July 1999. 2

[4] A. Lee, K. Pedersen, and D. Mumford, "The nonlinear statistics of high-contrast patches in natural images," *International Journal of Computer Vision*, vol. 54, no. 1, pp. 83–103, 2003. DOI: 10.1023/A:1023705401078. 2

[5] G. Carlsson *et al.*, "On the local behavior of spaces of natural images," *International Journal of Computer Vision*, vol. 2007, 2006. DOI: 10.1007/s11263-007-0056-x. 2

[6] D. Graham and D. Field, "Natural images: Coding efficiency," *Encyclopedia of Neuroscience*, vol. 6, pp. 19–27, 2008. 4

[7] S. Marčelja, "Mathematical description of the responses of simple cortical cells," *JOSA*, vol. 70, no. 11, pp. 1297–1300, 1980. DOI: 10.1364/JOSA.70.001297. 4

[8] J.P. Jones and L.A. Palmer, "An evaluation of the two-dimensional gabor filter model of simple receptive fields in cat striate cortex," *Journal of Neurophysiology*, vol. 58, no. 6, pp. 1233–1258, 1987. 4

[9] D. Field and D. Tolhurst, "The structure and symmetry of simple-cell receptive-field profiles in the cat's visual cortex," *Proceedings of the Royal society of London. Series B. Biological sciences*, vol. 228, no. 1253, pp. 379–400, 1986. 4

[10] D. Chandler and D. Field, "Estimates of the information content and dimensionality of natural scenes from proximity distributions," *JOSA A*, vol. 24, no. 4, pp. 922–941, 2007. DOI: 10.1364/JOSAA.24.000922. 4

[11] M. Riesenhuber and T. Poggio, "Neural mechanisms of object recognition," *Current opinion in neurobiology*, vol. 12, no. 2, pp. 162–168, 2002. DOI: 10.1016/S0959-4388(02)00304-5. 5

[12] D. J. Field, "What is the goal of sensory coding?," *Neural Computation*, vol. 6, pp. 559–601, 1994. DOI: 10.1162/neco.1994.6.4.559. 5

[13] B. A. Olshausen and D. J. Field, "Sparse coding with an overcomplete basis set: A strategy employed by v1?," *Vision Research*, vol. 37, no. 23, pp. 3311–3325, December 1997. DOI: 10.1016/S0042-6989(97)00169-7. 6, 29

[14] M. W. Marcellin, M. J. Gormish, A. Bilgin, and M. P. Boliek, "An overview of jpeg-2000 (2000)," in *Proceedings of IEEE Data Compression Conference*, Snowbird, Utah, 2000, pp. 523–541. 7

[15] M. Crouse, R. Nowak, and R. Baraniuk, "Wavelet-based statistical signal processing using hidden markov models," *IEEE Transactions on Signal Processing*, vol. 46, no. 4, pp. 886–902, April 1998. DOI: 10.1109/78.668544. 7

[16] K. N. Ramamurthy, J. J. Thiagarajan, and A. Spanias, "Template learning using wavelet domain statistical models," in *SGAI Conf.*, 2009, pp. 179–192. DOI: 10.1007/978-1-84882-983-1_13. 7

[17] R. Baraniuk, V. Cevher, M. Duarte, and C. Hegde, "Model-based compressive sensing," *IEEE Transactions on Information Theory*, vol. 56, no. 4, pp. 20, 2008. DOI: 10.1109/TIT.2010.2040894. 7

[18] L. He, H. Chen, and L. Carin, "Tree-structured compressive sensing with variational bayesian analysis," *IEEE Signal Processing Letters*, vol. 17, no. 3, pp. 233–236, 2010. DOI: 10.1109/LSP.2009.2037532. 7

[19] J. Shapiro, "Embedded image coding using zerotrees of wavelet coefficients," *IEEE Transactions on Signal Processing*, vol. 41, no. 12, pp. 3445–3462, December 1993. DOI: 10.1109/78.258085. 7

[20] M. Elad, *Sparse and Redundant Representations: From Theory to Applications in Signal and Image Processing*, Springer, 2010. 7, 30, 36

[21] Y. Eldar, P. Kuppinger, and H. Bolcskei, "Block-sparse signals: Uncertainty relations and efficient recovery," *IEEE Transactions on Signal Processing*, vol. 58, no. 6, pp. 3042–3054, 2010. DOI: 10.1109/TSP.2010.2044837. 7

[22] J. Huang, T. Zhang, and D. Metaxas, "Learning with structured sparsity," *Proceedings of the 26th Annual International Conference on Machine Learning ICML 09*, pp. 1–8, 2009. DOI: 10.1145/1553374.1553429. 7

[23] J. J. Thiagarajan, K. N. Ramamurthy, and A. Spanias, "Multilevel dictionary learning for sparse representation of images," in *Proceedings of the IEEE DSP Workshop, Sedona*, 2011. DOI: 10.1109/DSP-SPE.2011.5739224. 7

[24] M. Aharon and M. Elad, "Image denoising via sparse and redundant representations over learned dictionaries," *IEEE Transactions on Image Processing*, vol. 15, no. 12, pp. 3736–3745, 2006. DOI: 10.1109/TIP.2006.881969. 8, 37

[25] K. Dabov, A. Foi, V. Katkovnik, K. Egiazarian, et al., "Image denoising with block-matching and 3 d filtering," in *Proceedings of SPIE*, 2006, vol. 6064, pp. 354–365. 8

[26] E. Candes and M. Wakin, "An introduction to compressive sampling," *IEEE Signal Processing Magazine*, vol. 25, no. 2, pp. 21–30, 2008. DOI: 10.1109/MSP.2007.914731. 8

[27] S. Howard, A. Calderbank, and S. Searle, "A fast reconstruction algorithm for deterministic compressive sensing using second order reed-muller codes," *2008 42nd Annual Conference on Information Sciences and Systems*, pp. 11–15, 2008. DOI: 10.1109/CISS.2008.4558486. 8

[28] J.M. Duarte-Carvajalino and G. Sapiro, "Learning to sense sparse signals: Simultaneous sensing matrix and sparsifying dictionary optimization," *Image Processing, IEEE Transactions on*, vol. 18, no. 7, pp. 1395–1408, 2009. DOI: 10.1109/TIP.2009.2022459. 8, 58

[29] G.E. Hinton, "Learning to represent visual input," *Philosophical Transactions of the Royal Society B: Biological Sciences*, vol. 365, no. 1537, pp. 177–184, 2010. DOI: 10.1098/rstb.2009.0200. 9

[30] H. Lee, C. Ekanadham, and A. Ng, "Sparse deep belief net model for visual area V2," *Advances in neural information processing systems*, vol. 20, pp. 873–880, 2008. 9

[31] C. Poultney, S. Chopra, and Y. LeCun, "Efficient learning of sparse representations with an energy-based model," *Advances in neural information processing systems*, vol. 19, pp. 1137–1144, 2006. 9

[32] R. Rigamonti, M.A. Brown, and V. Lepetit, "Are sparse representations really relevant for image classification?," in *Computer Vision and Pattern Recognition (CVPR), 2011 IEEE Conference on*. IEEE, 2011, pp. 1545–1552. DOI: 10.1109/CVPR.2011.5995313. 9

[33] M. Brown, G. Hua, and S. Winder, "Discriminative learning of local image descriptors," *Pattern Analysis and Machine Intelligence, IEEE Transactions on*, vol. 33, no. 1, pp. 43–57, 2011. DOI: 10.1109/TPAMI.2010.54. 9

[34] T. Serre, L. Wolf, S. Bileschi, M. Riesenhuber, and T. Poggio, "Robust object recognition with cortex-like mechanisms," *Pattern Analysis and Machine Intelligence, IEEE Transactions on*, vol. 29, no. 3, pp. 411–426, 2007. DOI: 10.1109/TPAMI.2007.56. 9

[35] J. Mutch and D.G. Lowe, "Object class recognition and localization using sparse features with limited receptive fields," *International Journal of Computer Vision*, vol. 80, no. 1, pp. 45–57, 2008. DOI: 10.1007/s11263-007-0118-0. 10

[36] J. Mairal, F. Bach, J. Ponce, G. Sapiro, and A. Zisserman, "Supervised dictionary learning," in *Proceedings of NIPS*, 2009. 11, 71

[37] J. Wright *et.al.*, "Robust face recognition via sparse representation," *IEEE Trans. on PAMI*, vol. 31, no. 2, pp. 210–227, 2001. DOI: 10.1109/TPAMI.2008.79. 11, 72

[38] R. Raina *et.al.*, "Self-taught learning: Transfer learning from unlabeled data," in *Proceedings of ICML*, 2007. DOI: 10.1145/1273496.1273592. 11, 37, 73

[39] K. Yu *et.al.*, "Nonlinear learning using local coordinate coding," *Proceedings of NIPS*, 2009. 11, 12, 47, 78

[40] S. Lazebnik *et.al.*, "Beyond bags of features: Spatial pyramid matching for recognizing natural scene categories," *Proceedings of CVPR*, 2006. DOI: 10.1109/CVPR.2006.68. 11, 74

[41] R. Bracewell, *The Fourier Transform and Its Applications (3rd edition)*, McGraw-Hill Science Engineering, 1999. 13

[42] I. Daubechies, *Ten Lectures on Wavelets*, SIAM, 1992. DOI: 10.1137/1.9781611970104. 13

[43] E. J. Candès and D. L. Donoho, "New tight frames of curvelets and optimal representations of objects with $C^2$ singularities," Tech. Rep., Department of Statistics, Stanford University, USA, 2002. 13

[44] B. D. Rao and Y. Bresler, "Signal processing with sparseness constraints," in *Proceedings of the 1998 IEEE International Conference on Acoustics, Speech and Signal Processing*, Seattle, 1998. DOI: 10.1109/ICASSP.1998.681826. 13

[45] P. Frossard and P. Vandergheynst, "Redundant representations in image processing," in *Proceedings of the 2003 IEEE International Conference on Image Processing*, Barcelona, Spain, 2003. 13

[46] R. R. Coifman and D. L. Donoho, "Translational invariant de-noising," Tech. Rep., Wavelets and Statistics, Lecture Notes in Statistics, 1995. 13

[47] H. Zou and T. Hastie, "Regularization and variable selection via the elastic net," *Journal of the Royal Statistical Society: Series B (Statistical Methodology)*, vol. 67, no. 2, pp. 301–320, 2005. DOI: 10.1111/j.1467-9868.2005.00503.x. 15

[48] M. Yuan and Y. Lin, "Model selection and estimation in regression with grouped variables," *Journal of the Royal Statistical Society: Series B (Statistical Methodology)*, vol. 68, no. 1, pp. 49–67, 2005. DOI: 10.1111/j.1467-9868.2005.00532.x. 15

[49] J. Friedman, T. Hastie, and R. Tibshirani, "A note on the group lasso and a sparse group lasso," *arXiv preprint arXiv:1001.0736*, 2010. 15

[50] J. J. Thiagarajan K. N. Ramamurthy and A. Spanias, "Improved sparse coding using manifold projections," in *IEEE ICIP*, 2011. DOI: 10.1109/ICIP.2011.6115656. 16, 48

[51] J.F. Gemmeke, T. Virtanen, and A. Hurmalainen, "Exemplar-based sparse representations for noise robust automatic speech recognition," *IEEE Transactions on Audio, Speech, and Language Processing*, vol. 19, no. 7, pp. 2067–2080, sept. 2011. DOI: 10.1109/TASL.2011.2112350. 16

[52] M. Slawski and M. Hein, "Sparse recovery for protein mass spectrometry data," in *NIPS Workshop on Practical Application of Sparse Modeling: Open Issues and New Directions*, 2010. 16

[53] JM Bardsley, "Covariance-preconditioned iterative methods for nonnegatively constrained astronomical imaging," *SIAM journal on matrix analysis and applications*, vol. 27, no. 4, pp. 1184–1197, 2006. DOI: 10.1137/040615043. 16

[54] D. Donoho, I. Johnstone, and J. Hoch, "Maximum entropy and the nearly black object," *Journal of the Royal Statistical Society*, vol. 54, no. 1, pp. 41–81, 1992. 16

[55] L. Benaroya, L.M. Donagh, F. Bimbot, and R. Gribonval, "Non negative sparse representation for Wiener based source separation with a single sensor," *IEEE ICASSP*, vol. 1, pp. VI–613–16, 2003. DOI: 10.1109/ICASSP.2003.1201756. 16

[56] B. Cheng, J. Yang, S. Yan, Y. Fu, and T. Huang, "Learning with l1-graph for image analysis," *IEEE transactions on image processing*, vol. 19, no. 4, pp. 858–66, Apr. 2010. DOI: 10.1109/TIP.2009.2038764. 16

[57] R. He and W. Zheng, "Nonnegative sparse coding for discriminative semi-supervised learning," in *IEEE Conf. on Computer Vision and Pattern Recognition*, 2011. DOI: 10.1109/CVPR.2011.5995487. 16

[58] D. L. Donoho and J. Tanner, "Counting the faces of randomly-projected hypercubes and orthants, with applications," *Discrete & computational geometry*, vol. 43, Apr. 2010. DOI: 10.1007/s00454-009-9221-z. 16, 20

[59] M. Wang, W. Xu, and A. Tang, "A unique nonnegative solution to an underdetermined system: From vectors to matrices," *IEEE Transactions on Signal Processing*, vol. 59, no. 3, pp. 1007–1016, march 2011. DOI: 10.1109/TSP.2010.2089624. 16

[60] D. L. Donoho and M. Elad, "Optimally sparse representation in general (nonorthogonal) dictionaries via $l^1$ minimization," *Proceedings of the National Academy of Sciences of the United States of America*, vol. 100, no. 5, pp. 2197–2202, March 2003. DOI: 10.1073/pnas.0437847100. 17, 19

[61] D. Donoho, "Neighborly polytopes and sparse solution of underdetermined linear equations," Tech. Rep., Stanford University, 2005. 17

[62] J.A. Tropp, "On the conditioning of random subdictionaries," *Applied and Computational Harmonic Analysis*, vol. 25, no. 1, pp. 1–24, 2008. DOI: 10.1016/j.acha.2007.09.001. 19

[63] D.L. Donoho and J. Tanner, "Precise undersampling theorems," *Proceedings of the IEEE*, vol. 98, no. 6, pp. 913–924, 2010. DOI: 10.1109/JPROC.2010.2045630. 20, 58

[64] D. L. Donoho and J. Tanner, "Sparse nonnegative solution of underdetermined linear equations by linear programming," *Proceedings of the National Academy of Sciences*, vol. 102, no. 3, Jun. 2005. DOI: 10.1073/pnas.0502269102. 20

[65] D.L. Donoho, "Neighborly polytopes and sparse solutions of underdetermined linear equations," Tech. Rep., Stanford University, 2005. 20

[66] J.D. Blanchard, C. Cartis, J. Tanner, and A. Thompson, "Phase transitions for greedy sparse approximation algorithms," *Applied and Computational Harmonic Analysis*, vol. 30, no. 2, pp. 188–203, 2011. DOI: 10.1016/j.acha.2010.07.001. 20

[67] A. Maleki and D.L. Donoho, "Optimally tuned iterative reconstruction algorithms for compressed sensing," *Selected Topics in Signal Processing, IEEE Journal of*, vol. 4, no. 2, pp. 330–341, 2010. DOI: 10.1109/JSTSP.2009.2039176. 20

[68] G. Davis, S. Mallat, and M. Avellaneda, "Greedy adaptive approximation," *Journal of Constructive Approximation*, vol. 13, pp. 57–98, 1997. DOI: 10.1007/s003659900033. 20

[69] S. Mallat and Z. Zhang, "Matching pursuits with time-frequency dictionaries," *IEEE Transactions on Signal Processing*, vol. 41, no. 12, pp. 3397–3415, 1993. DOI: 10.1109/78.258082. 20, 23

[70] J. A. Tropp, "Greed is good: Algorithmic results for sparse approximation," *IEEE Transactions on Information Theory*, vol. 50, no. 10, pp. 2231–2242, October 2004. DOI: 10.1109/TIT.2004.834793. 20, 21, 23

[71] S. S. Chen, D. L. Donoho, and M. A. Saunders, "Atomic decomposition by basis pursuit," *SIAM Review*, vol. 43, no. 1, pp. 129–159, 2001. DOI: 10.1137/S003614450037906X. 20, 22, 23

[72] I. F. Gorodnitsky and B. D. Rao, "Sparse signal reconstruction from limited data using FOCUSS: A re-weighted norm minimization algorithm," *IEEE Transactions on Signal Processing*, vol. 45, no. 3, pp. 600–616, March 1997. DOI: 10.1109/78.558475. 20

[73] H. Lee, A. Battle, R. Raina, and A.Y. Ng, "Efficient sparse coding algorithms," *Advances in neural information processing systems*, vol. 19, pp. 801, 2007. 20, 28

[74] B. Efron, T. Hastie, I. Johnstone, and R. Tibshirani, "Least angle regression," *The Annals of statistics*, vol. 32, no. 2, pp. 407–499, 2004. DOI: 10.1214/009053604000000067. 20, 50

[75] M. Elad, "Why simple shrinkage is still relevant for redundant representations?," *IEEE Transactions on Information Theory*, vol. 52, no. 12, pp. 5559–5569, December 2006. DOI: 10.1109/TIT.2006.885522. 20

[76] M. Elad, B. Matalon, J. Shtok, and M. Zibulevsky, "A wide-angle view at iterated shrinkage algorithms," in *SPIE (Wavelet XII) 2007*, 2007. DOI: 10.1117/12.741299. 20

[77] J. A. Tropp and A. Gilbert, "Signal recovery from partial information via orthogonal matching pursuit," *IEEE Transactions on Information Theory*, 2005. 21

[78] J.J. Fuchs, "Recovery of exact sparse representations in the presence of bounded noise," *Information Theory, IEEE Transactions on*, vol. 51, no. 10, pp. 3601–3608, 2005. DOI: 10.1109/TIT.2005.855614. 21

[79] S. Boyd and L. Vandenberghe, *Convex Optimization*, Cambridge University Press, 2004. DOI: 10.1017/CBO9780511804441. 22

[80] L. K. Jones, "On a conjecture of huber concerning the convergence of projection pursuit regression," *Annals of Statistics*, vol. 15, no. 2, pp. 880–882, 1987. DOI: 10.1214/aos/1176350382. 23

[81] S. Chen, S. A. Billings, , and W. Luo, "Orthogonal least squares methods and their application to nonlinear system identification," *International Journal of Control*, vol. 50, no. 5, pp. 1873–1896, 1989. DOI: 10.1080/00207178908953472. 24

[82] G. Davis, S. Mallat, and Z. Zhang, "Adaptive time-frequency decompositions," *Optical Engineering*, vol. 33, no. 7, pp. 2183–2191, July 1994. DOI: 10.1117/12.173207. 24

[83] Y. C. Pati, R. Rezaiifar, and P. S. Krishnaprasad, "Orthogonal matching pursuit: Recursive function approximation with applications to wavelet decomposition," in *Proceedings of 27th Annual Asilomar Conference on Signals, Systems and Computers*, Pacific Grove, California, November 1993. DOI: 10.1109/ACSSC.1993.342465. 24

[84] S. Weisberg, *Applied linear regression*, vol. 528, Wiley, 2005. DOI: 10.1002/0471704091. 24

[85] A.M. Bruckstein, Michael Elad, and Michael Zibulevsky, "On the uniqueness of nonnegative sparse solutions to underdetermined systems of equations," *IEEE Transactions on Information Theory*, vol. 54, no. 11, pp. 4813–4820, 2008. DOI: 10.1109/TIT.2008.929920. 25

[86] D. Needell and J.A. Tropp, "Cosamp: Iterative signal recovery from incomplete and inaccurate samples," *Applied and Computational Harmonic Analysis*, vol. 26, no. 3, pp. 301–321, 2009. DOI: 10.1016/j.acha.2008.07.002. 25

[87] J. A. Tropp, "Algorithms for simultaneous sparse approximation. Part I: Greedy pursuit," *Signal Processing*, vol. 86, pp. 572–588, April 2006. DOI: 10.1016/j.sigpro.2005.05.030. 25

[88] D. Donoho, "Wavelet shrinkage and w.v.d.: A 10-minute tour," in *Progress in Wavelet Analysis and Applications*, 1993, pp. 109–128. 28

[89] M. Elad, B. Matalon, J. Shtok, and M. Zibulevsky, "A wide-angle view at iterated shrinkage algorithms," in *in SPIE (Wavelet XII*, 2007, pp. 26–29. DOI: 10.1117/12.741299. 28

[90] J.L. Starck, F. Murtagh, and J.M. Fadili, *Sparse image and signal processing: wavelets, curvelets, morphological diversity*, Cambridge Univ Pr, 2010. DOI: 10.1017/CBO9780511730344. 29

[91] J.L. Starck, D.L. Donoho, and E.J. Candes, "Very high quality image restoration by combining wavelets and curvelets," in *Proc. SPIE*, 2001, vol. 4478, pp. 9–19. DOI: 10.1117/12.449693. 29

[92] R. Gribonval and M. Nielsen, "Sparse representations in unions of bases," *IEEE Transactions on Information Theory*, vol. 49, no. 12, pp. 3320–3325, 2003. DOI: 10.1109/TIT.2003.820031. 29

[93] S. Mallat, *A wavelet tour of signal processing*, Academic press, 1999. 29

[94] B. Olshausen, "Sparsenet matlab toolbox," Available at http://redwood.berkeley.edu/bruno/sparsenet/. 30

[95] M. Aharon, M. Elad, and A. Bruckstein, "K-SVD: An algorithm for designing overcomplete dictionaries for sparse representation," *IEEE Transactions on Signal Processing*, vol. 54, no. 11, pp. 4311–4322, 2006. DOI: 10.1109/TSP.2006.881199. 30, 34, 36

[96] K. Engan, S. O. Aase, and J. H. Husoy, "Method of optimal directions for frame design," in *Proceedings of IEEE ICASSP*, 1999. DOI: 10.1109/ICASSP.1999.760624. 30, 33

[97] K. Engan, B.D. Rao, and K. Kreutz-Delgado, "Frame design using FOCUSS with method of optimal directions (MoD)," in *Proceedings of Norwegian Signal Processing Symposium*, 1999. 30, 33

[98] Z. He *et.al.*, "K-hyperline clustering learning for sparse component analysis," *Signal Processing*, vol. 89, pp. 1011–1022, 2009. DOI: 10.1016/j.sigpro.2008.12.005. 31

[99] J.J. Thiagarajan, K.N. Ramamurthy, and A. Spanias, "Optimality and stability of the $k$–hyperline clustering algorithm," *Pattern Recognition Letters*, vol. 32, no. 9, pp. 1299–1304, 2011. DOI: 10.1016/j.patrec.2011.03.005. 32, 51

[100] I. F. Gorodnitsky and B. D. Rao, "Sparse signal reconstruction from limited data using focus: A re-weighted norm minimization algorithm," *IEEE Transactions on Signal Processing*, vol. 45, no. 3, pp. 600–616, March 1997. DOI: 10.1109/78.558475. 33

[101] S. F. Cotter, M. N. Murthi, and B. D. Rao, "Fast basis selection methods," in *Proceedings of 31st Annual Asilomar Conference on Signals, Systems and Computers*, Pacific Grove, California, November 1997, vol. 2. DOI: 10.1109/ACSSC.1997.679149. 33

[102] J.M. Duarte-Carvajalino *et.al.*, "Learning to sense sparse signals: simultaneous sensing matrix and sparsifying dictionary optimization," *IEEE Transactions on Image Processing*, vol. 18, no. 7, pp. 1395–1408, 2009. DOI: 10.1109/TIP.2009.2022459. 37

[103] S. Zhu *et.al.*, "Learning explicit and implicit visual manifolds by information projection," *Pattern Recognition Letters*, vol. 31, pp. 667–685, 2010. DOI: 10.1016/j.patrec.2009.07.020. 37

[104] J. Mairal, F. Bach, J. Ponce, and G. Sapiro, "Online dictionary learning for sparse coding," in *Proceedings of the 26th Annual International Conference on Machine Learning*. ACM, 2009, pp. 689–696. DOI: 10.1145/1553374.1553463. 42

[105] L. Bottou and O. Bousquet, "The tradeoffs of large-scale learning," *Optimization for Machine Learning*, p. 351, 2011. 42

[106] G. Yu, G. Sapiro, and S. Mallat, "Image modeling and enhancement via structured sparse model selection," in *Proc. of IEEE ICIP*, Sep. 2010, pp. 1641–1644. DOI: 10.1109/ICIP.2010.5653853. 43

[107] J. Mairal, F. Bach, J. Ponce, G. Sapiro, and A. Zisserman, "Non-local sparse models for image restoration," in *Computer Vision, 2009 IEEE 12th International Conference on*. IEEE, 2009, pp. 2272–2279. DOI: 10.1109/ICCV.2009.5459452. 43, 44

[108] S. Roweis and L. Saul, "Nonlinear dimensionality reduction by locally linear embedding," *Science*, pp. 2323–2326, 2000. DOI: 10.1126/science.290.5500.2323. 47

[109] K. N. Ramamurthy, J. J. Thiagarajan, and A. Spanias, "Recovering non-negative and combined sparse representations," *arXiv preprint arXiv:1303.4694*, 2013. DOI: 10.1016/j.dsp.2013.11.003. 50

[110] A. Caponnetto and A. Rakhlin, "Stability properties of empirical risk minimization over Donsker classes," *Journal of Machine Learning Research*, vol. 7, pp. 2565–2583, 2006. 50

[111] T. Poggio, R. Rifkin, S. Mukherjee, and P. Niyogi, "General conditions for predictivity in learning theory," *Nature*, vol. 428, no. 6981, pp. 419–422, 2004. DOI: 10.1038/nature02341. 50

[112] A. Rakhlin and A. Caponnetto, "Stability of K-means clustering," in *Advances in Neural Information Processing Systems*, B. Schölkopf, J. Platt, , and T. Hoffman, Eds., Cambridge, MA, 2007, vol. 19, MIT Press. 51

[113] S. Ben-David, U. von Luxburg, and D. Pál, "A sober look at clustering stability," *Proceedings of the Conference on Computational Learning Theory*, pp. 5–19, 2006. DOI: 10.1007/11776420_4. 51

[114] S. Ben-David, D. Pál, and Hans Ulrich Simon, "Stability of K-means clustering," 2007, vol. 4539 of *Lecture Notes in Computer Science*, pp. 20–34, Springer. 51

[115] A Maurer and M Pontil, "$k$-dimensional coding schemes in Hilbert spaces," vol. 56, no. 11, pp. 5839–5846, 2010. 51

[116] D. Vainsencher and A. Bruckstein, "The Sample Complexity of Dictionary Learning," *Journal of Machine Learning Research*, vol. 12, pp. 3259–3281, 2011. 51

[117] J. Buhmann, "Empirical risk approximation: An induction principle for unsupervised learning," Tech. Rep. IAI-TR-98-3, The University of Bonn, 1998. 51

[118] J. J. Thiagarajan, K. N. Ramamurthy, and A. Spanias, "Learning stable multilevel dictionaries for sparse representation of images," *IEEE Transactions on Neural Networks and Learning Systems (arxiv.org/pdf/1303.0448v1)*, 2013. 52

[119] D.L. Donoho, "Compressed sensing," *Information Theory, IEEE Transactions on*, vol. 52, no. 4, pp. 1289–1306, 2006. DOI: 10.1109/TIT.2006.871582. 55, 57

[120] E.J. Candès and M.B. Wakin, "An introduction to compressive sampling," *Signal Processing Magazine, IEEE*, vol. 25, no. 2, pp. 21–30, 2008. DOI: 10.1109/MSP.2007.914731. 55, 57

[121] J. Romberg, "Imaging via compressive sampling," *Signal Processing Magazine, IEEE*, vol. 25, no. 2, pp. 14–20, 2008. DOI: 10.1109/MSP.2007.914729. 56

[122] E.J. Candès, J. Romberg, and T. Tao, "Robust uncertainty principles: Exact signal reconstruction from highly incomplete frequency information," *Information Theory, IEEE Transactions on*, vol. 52, no. 2, pp. 489–509, 2006. DOI: 10.1109/TIT.2005.862083. 57

[123] T.T. Cai, L. Wang, and G. Xu, "Stable recovery of sparse signals and an oracle inequality," *Information Theory, IEEE Transactions on*, vol. 56, no. 7, pp. 3516–3522, 2010. DOI: 10.1109/TIT.2010.2048506. 57

[124] W.B. Johnson and J. Lindenstrauss, "Extensions of lipschitz mappings into a hilbert space," *Contemporary mathematics*, vol. 26, no. 189-206, pp. 1, 1984. DOI: 10.1090/conm/026/737400. 57

[125] R. Baraniuk, M. Davenport, R. DeVore, and M. Wakin, "A simple proof of the restricted isometry property for random matrices," *Constructive Approximation*, vol. 28, no. 3, pp. 253–263, 2008. DOI: 10.1007/s00365-007-9003-x. 57, 58

[126] A. Cohen, W. Dahmen, and R. DeVore, "Compressed sensing and best k-term approximation," *J. Amer. Math. Soc*, vol. 22, no. 1, pp. 211–231, 2009. DOI: 10.1090/S0894-0347-08-00610-3. 58

[127] M. Elad, "Optimized projections for compressed sensing," *Signal Processing, IEEE Transactions on*, vol. 55, no. 12, pp. 5695–5702, 2007. DOI: 10.1109/TSP.2007.900760. 58

[128] M. F. Duarte, M. A. Davenport, D. Takhar, J. N. Laska, T. Sun, K. F. Kelly, and R. G. Baraniuk, "Single-pixel imaging via compressive sampling," *IEEE Signal Process. Mag.*, vol. 25, no. 2, pp. 83–91, 2008. DOI: 10.1109/MSP.2007.914730. 61

[129] M. B. Wakin, J. N. Laska, M. F. Duarte, D. Baron, S. Sarvotham, D. Takhar, K. F. Kelly, and R. G. Baraniuk, "Compressive imaging for video representation and coding," in *Pict. Coding Symp.*, Apr. 2006. 61

[130] A. Wagadarikar, R. John, R. Willett, and D. Brady, "Single disperser design for coded aperture snapshot spectral imaging," *App. Optics*, vol. 47, no. 10, pp. 44–51, 2008. DOI: 10.1364/AO.47.000B44. 61

[131] N. Vaswani, "Kalman filtered compressed sensing," in *IEEE Conf. Image Process.*, Oct. 2008. DOI: 10.1109/ICIP.2008.4711899. 61

[132] N. Vaswani and W. Lu, "Modified-cs: Modifying compressive sensing for problems with partially known support," in *Intl. Symp. Inf. Theory*, June 2009. DOI: 10.1109/ISIT.2009.5205717. 61

[133] A. Veeraraghavan, D. Reddy, and R. Raskar, "Coded strobing photography: Compressive sensing of high speed periodic events," *IEEE Trans. Pattern Anal. Mach. Intell.*, vol. 33, no. 4, pp. 671–686, 2011. DOI: 10.1109/TPAMI.2010.87. 62

[134] G. Doretto, A. Chiuso, Y. N. Wu, and S. Soatto, "Dynamic textures," *Intl. J. Comp. Vision*, vol. 51, no. 2, pp. 91–109, 2003. DOI: 10.1023/A:1021669406132. 62

[135] A. B. Chan and N. Vasconcelos, "Probabilistic kernels for the classification of auto-regressive visual processes," in *IEEE Conf. Comp. Vision and Pattern Recog*, June 2005. DOI: 10.1109/CVPR.2005.279. 62, 65

[136] A. Veeraraghavan, A. Roy-Chowdhury, and R. Chellappa, "Matching shape sequences in video with an application to human movement analysis," *IEEE Trans. Pattern Anal. Mach. Intell.*, vol. 27, no. 12, pp. 1896–1909, 2005. DOI: 10.1109/TPAMI.2005.246. 62, 63

[137] P. Turaga, A. Veeraraghavan, and R. Chellappa, "Unsupervised view and rate invariant clustering of video sequences," *Comp. Vision and Image Understd.*, vol. 113, no. 3, pp. 353–371, 2009. DOI: 10.1016/j.cviu.2008.08.009. 62

[138] A. C. Sankaranarayanan, P. K. Turaga, R. G. Baraniuk, and R. Chellappa, "Compressive acquisition of dynamic scenes," in *ECCV (1)*, 2010, pp. 129–142. DOI: 10.1007/978-3-642-15549-9_10. 62

[139] N. Dalal and B. Triggs, "Histograms of oriented gradients for human detection," in *IEEE Conf. Comp. Vision and Pattern Recog*, 2005, pp. 886–893. DOI: 10.1109/CVPR.2005.177. 63

[140] R. Chaudhry, A. Ravichandran, G. D. Hager, and R. Vidal, "Histograms of oriented optical flow and binet-cauchy kernels on nonlinear dynamical systems for the recognition of human actions," in *IEEE Conf. Comp. Vision and Pattern Recog*, 2009, pp. 1932–1939. DOI: 10.1109/CVPR.2009.5206821. 63

[141] I. Laptev and T. Lindeberg, "Space-time interest points," *IEEE Intl. Conf. Comp. Vision.*, 2003. DOI: 10.1109/ICCV.2003.1238378. 63

[142] V. Cevher, A. C. Sankaranarayanan, M. F. Duarte, D. Reddy, R. G. Baraniuk, and R. Chellappa, "Compressive sensing for background subtraction," in *Euro. Conf. Comp. Vision*, Oct. 2008. DOI: 10.1007/978-3-540-88688-4_12. 63

[143] K. Kulkarni and P. Turaga, "Recurrence textures for human activity recognition from compressive cameras," in *IEEE Conf. on Image Processing (ICIP)*, 2012. DOI: 10.1109/ICIP.2012.6467135. 63, 64

[144] J. P. Eckmann, S. O. Kamphorst, and D. Ruelle, "Recurrence plots of dynamical systems," *Europhysics Letters*, vol. 5, no. 9, pp. 973–977, 1987. DOI: 10.1209/0295-5075/4/9/004. 64

[145] K. Fukunaga, *Introduction to statistical pattern recognition (2nd ed.)*, Academic Press, 1990. 71

[146] Ke Huang and Selin Aviyente, "Sparse representation for signal classification," *East*, vol. 19, pp. 609–616, 2007. 71

[147] F. Rodriguez and G. Sapiro, "Sparse representations for image classification: Learning discriminative and reconstructive non-parametric dictionaries," *Security*, 2008. 71

[148] J. J. Thiagarajan *et.al.*, "Sparse representations for automatic target classification in sar images," in *Proceedings of ISCCSP*, 2010. DOI: 10.1109/ISCCSP.2010.5463416. 72

[149] J. Yang *et.al.*, "Linear spatial pyramid matching using sparse coding for image classification," in *Proceedings of IEEE CVPR*, 2009. DOI: 10.1109/CVPR.2009.5206757. 75

[150] J. Yang, K. Yu, and T. Huang, "Supervised translation-invariant sparse coding.," in *CVPR'10*, 2010, pp. 3517–3524. DOI: 10.1109/CVPR.2010.5539958. 77

[151] L. Cayton, "Algorithms for manifold learning," Technical report, University of California, 2005. 78

[152] J. J. Thiagarajan and A. Spanias, "Learning dictionaries for local sparse coding in image classification," in *Proceedings of Asilomar SSC*, 2011. 78, 79, 80

[153] J. Wang *et.al.*, "Locality-constrained linear coding for image classification," in *Proceedings of IEEE CVPR*, 2010. DOI: 10.1109/CVPR.2010.5540018. 79

[154] H.T. Chen, H.W. Chang, and T.L. Liu, "Local discriminant embedding and its variants," in *IEEE CVPR*, 2005, vol. 2, pp. 846–853. DOI: 10.1109/CVPR.2005.216. 80, 82

[155] D. Cai, X. He, and J. Han, "Semi-supervised discriminant analysis," in *IEEE ICCV*, 2007, pp. 1–7. DOI: 10.1109/ICCV.2007.4408856. 81, 83

[156] S. Gao, I.W.H. Tsang, L.T. Chia, and P. Zhao, "Local features are not lonely–Laplacian sparse coding for image classification," in *IEEE CVPR*, 2010, pp. 3555–3561. DOI: 10.1109/CVPR.2010.5539943. 81

[157] K.N. Ramamurthy, J.J. Thiagarajan, P. Sattigeri, and A. Spanias, "Learning dictionaries with graph embedding constraints," in *Proceedings of Asilomar SSC*, 2012. DOI: 10.1109/ACSSC.2012.6489385. 81, 83

[158] N. Cristianini and J. Shawe-Taylor, *An introduction to support Vector Machines: and other kernel–based learning methods*, Cambridge University Press, 2000. DOI: 10.1017/CBO9780511801389. 84

[159] S. Gao, I. Tsang, and L.T. Chia, "Kernel sparse representation for image classification and face recognition," *Computer Vision–ECCV 2010*, pp. 1–14, 2010. DOI: 10.1007/978-3-642-15561-1_1. 84, 85

[160] H. V. Nguyen *et. al.*, "Kernel dictionary learning," in *Proceedings of the IEEE ICASSP*, 2012. 85

[161] J.T.Y. Kwok and I.W.H. Tsang, "The pre-image problem in kernel methods," *Neural Networks, IEEE Transactions on*, vol. 15, no. 6, pp. 1517–1525, 2004. DOI: 10.1109/TNN.2004.837781. 86

[162] J.J. Thiagarajan, D. Rajan, K.N. Ramamurthy, D. Frakes, and A. Spanias, "Automatic tumor identification using kernel sparse representations," in *Proceedings of IEEE BIBE*, 2012. DOI: 10.1109/BIBE.2012.6399658. 89

# Authors' Biographies

## JAYARAMAN J. THIAGARAJAN

**Jayaraman J. Thiagarajan** received his M.S. and Ph.D. degrees in Electrical Engineering from Arizona State University. He is currently a postdoctoral researcher in the Center for Applied Scientific Computing at Lawrence Livermore National Laboratory. His research interests are in the areas of machine learning, computer vision, and data analysis and visualization. He has served as a reviewer for several IEEE, Elsevier, and Springer journals and conferences.

## KARTHIKEYAN NATESAN RAMAMURTHY

**Karthikeyan Natesan Ramamurthy** is a research staff member in the Business Solutions and Mathematical Sciences department at the IBM Thomas J. Watson Research Center in Yorktown Heights, NY. He received his M.S. and Ph.D. degrees in Electrical Engineering from Arizona State University. His research interests are in the areas of low-dimensional signal models, machine learning, data analytics, and computer vision. He has been a reviewer for a number of IEEE and Elsevier journals and conferences.

## PAVAN TURAGA

**Pavan Turaga** is an Assistant Professor with the School of Arts, Media, and Engineering and the School of Electrical, Computer, and Energy Engineering at Arizona State University, since 2011. Prior to that, he was a Research Associate at the Center for Automation Research, University of Maryland, College Park, MD, from 2009-11. He received M.S. and Ph.D. degrees in Electrical Engineering from the University of Maryland, College Park, MD, in 2008 and 2009 respectively, and the B.Tech. degree in Electronics and Communication Engineering from the Indian Institute of Technology, Guwahati, India, in 2004. His research interests are in computer vision, applied statistics, and machine learning with applications to human activity analysis, video summarization, and dynamic scene analysis. He was awarded the Distinguished Dissertation Fellowship in 2009. He was selected to participate in the Emerging Leaders in Multimedia Workshop by IBM, New York, in 2008.

## ANDREAS SPANIAS

**Andreas Spanias** is Professor in the School of Electrical, Computer, and Energy Engineering at Arizona State University (ASU). He is also the founder and director of the SenSIP industry consortium. His research interests are in the areas of adaptive signal processing, speech processing, and audio sensing. He and his student team developed the computer simulation software Java-DSP (J-DSP - ISBN 0-9724984-0-0). He is author of two text books: *Audio Processing and Coding* by Wiley and *DSP: An Interactive Approach.* He served as Associate Editor of the *IEEE Transactions on Signal Processing* and as General Co-chair of IEEE ICASSP-99. He also served as the IEEE Signal Processing Vice-President for Conferences. Andreas Spanias is co-recipient of the 2002 IEEE Donald G. Fink paper prize award and was elected Fellow of the IEEE in 2003. He served as Distinguished lecturer for the IEEE Signal processing society in 2004.

Printed in the United States
by Baker & Taylor Publisher Services